Developing Essential Understanding of Statistics

for Teaching Mathematics in Grades 9–12

Patricia S. Wilson
Volume Editor
University of Georgia
Athens, Georgia

Rose Mary Zbiek
Series Editor
The Pennsylvania State University
University Park, Pennsylvania

Roxy Peck
California Polytechnic State University
San Luis Obispo, California

Robert Gould
University of California, Los Angeles
Los Angeles, California

Stephen J. Miller
Winchester Thurston School
Pittsburgh, Pennsylvania

NATIONAL COUNCIL OF
TEACHERS OF MATHEMATICS

www.nctm.org/more4u
Access code: EUS13804

Copyright © 2013 by
The National Council of Teachers of Mathematics, Inc.
1906 Association Drive, Reston, VA 20191-1502
(703) 620-9840; (800) 235-7566; www.nctm.org
All rights reserved

Library of Congress Cataloging-in-Publication Data

Peck, Roxy.
  Developing essential understanding of statistics for teaching mathematics in grades 9-12 / Roxy Peck, California Polytechnic State University, San Luis Obispo, California, Robert Gould, University of California-Los Angeles, Los Angeles, California, Stephen J. Miller, Winchester Thurston School, Pittsburgh, Pennsylvania; Patricia S. Wilson, volume editor, University of Georgia, Athens, Georgia; Rose Mary Zbiek, series editor, The Pennsylvania State University, University Park, Pennsylvania.
       pages cm
  Includes bibliographical references.
  ISBN 978-0-87353-676-9
  1. Mathematical statistics--Study and teaching (Secondary) 2. Effective teaching. I. Gould, Robert, 1965- author. II. Miller, Stephen J., 1966- author. III. Title.
  QA276.18.P425 2012
  519.5071'273--dc23
                        2012034200

The National Council of Teachers of Mathematics is the public voice of mathematics education, supporting teachers to ensure equitable mathematics learning of the highest quality for all students through vision, leadership, professional development, and research.

When forms, problems, or sample documents are included or are made available on NCTM's website, their use is authorized for educational purposes by educators and noncommercial or nonprofit entities that have purchased this book. Except for that use, permission to photocopy or use material electronically from *Developing Essential Understanding of Statistics for Teaching Mathematics in Grades 9–12* must be obtained from www.copyright.com or by contacting Copyright Clearance Center, Inc. (CCC), 222 Rosewood Drive, Danvers, MA 01923, 978-750-8400. CCC is a not-for-profit organization that provides licenses and registration for a variety of users. Permission does not automatically extend to any items identified as reprinted by permission of other publishers or copyright holders. Such items must be excluded unless separate permissions are obtained. It is the responsibility of the user to identify such materials and obtain the permissions.

The publications of the National Council of Teachers of Mathematics present a variety of viewpoints. The views expressed or implied in this publication, unless otherwise noted, should not be interpreted as official positions of the Council.

Printed in the United States of America

Contents

Foreword v

Preface vii

Introduction 1
    Why Statistics? 1
    Understanding Statistical Concepts 2
    Big Ideas and Essential Understandings 3
    Benefits for Teaching, Learning, and Assessing 4
    Ready to Begin 5

Chapter 1 7
Statistics: The Big Ideas and Essential Understandings
    Big Ideas and Associated Understandings 8
        The big ideas in brief 8
        The complete list of ideas and understandings 10
    The Nature of Statistical Models: Big Idea 1 13
        Mathematical models describe structure 14
        Statistical models include variation 17
        Good models give useful descriptions of structure 18
    Describing Variability: Big Idea 2 24
        Variability in a population 25
        Approximating population distributions 29
        Variability in the values of a sample statistic 32
        Approximating sampling distributions 34
    The Logic of Hypothesis Testing: Big Idea 3 44
        Choosing between a null and an alternative hypothesis 45
        Determining the alternative hypothesis 47
        Convincing evidence against the null hypothesis 49
        A measure of surprise 52
        A risk of error 54
    The Importance of the Data Collection Method: Big Idea 4 57
        Observational studies and controlled experiments 57
        Random assignment 61
        Random selection 64
        Implications of random selection and random assignment 65
    Evaluating Estimators: Big Idea 5 67
        The quality of estimators 68
        Potential bias in estimators 69
        Precision based on standard error and bias 71
        Confidence intervals as estimators 74
        Precision depends on sampling method and sample size 78

    Sample size 80
    Good methods trump large samples 83
    Small samples despite large populations 85
  Conclusion 87

Chapter 2 89
Connections: Looking Back and Ahead in Learning

  Differences between Statistics and Mathematics 89
  Informal to Semiformal to Formal Reasoning 90
  Reasoning Related to Each Big Idea 91
    The growth of Big Idea 1 91
    The growth of Big Idea 2 92
    The growth of Big Idea 3 93
    The growth of Big Idea 4 93
    The growth of Big Idea 5 94
  Conclusion 95

Chapter 3 97
Challenges: Learning, Teaching, and Assessing

  Going beyond Computational Fluency 97
  Producing Good Statistics Questions 99
  Assessing Interpretation and Conceptual Understanding 102
  The Data Analysis Process—Culminating Assessments 107
  Conclusion 107

Appendix 1 109
Glossary of Statistical Terms

Appendix 2 114
Resources for Teachers

References 120

Foreword

Teaching mathematics in prekindergarten–grade 12 requires a special understanding of mathematics. Effective teachers of mathematics think about and beyond the content that they teach, seeking explanations and making connections to other topics, both inside and outside mathematics. Students meet curriculum and achievement expectations when they work with teachers who know what mathematics is important for each topic that they teach.

The National Council of Teachers of Mathematics (NCTM) presents the Essential Understanding Series in tandem with a call to focus the school mathematics curriculum in the spirit of *Curriculum Focal Points for Prekindergarten through Grade 8 Mathematics: A Quest for Coherence*, published in 2006, and *Focus in High School Mathematics: Reasoning and Sense Making*, released in 2009. The Essential Understanding books are a resource for individual teachers and groups of colleagues interested in engaging in mathematical thinking to enrich and extend their own knowledge of particular mathematics topics in ways that benefit their work with students. The topic of each book is an area of mathematics that is difficult for students to learn, challenging to teach, and critical for students' success as learners and in their future lives and careers.

Drawing on their experiences as teachers, researchers, and mathematicians, the authors have identified the big ideas that are at the heart of each book's topic. A set of essential understandings—mathematical points that capture the essence of the topic—fleshes out each big idea. Taken collectively, the big ideas and essential understandings give a view of a mathematics that is focused, connected, and useful to teachers. Links to topics that students encounter earlier and later in school mathematics and to instruction and assessment practices illustrate the relevance and importance of a teacher's essential understanding of mathematics.

On behalf of the Board of Directors, I offer sincere thanks and appreciation to everyone who has helped to make this series possible. I extend special thanks to Rose Mary Zbiek for her leadership as series editor. I join the Essential Understanding project team in welcoming you to these books and in wishing you many years of continued enjoyment of learning and teaching mathematics.

Henry Kepner
President, 2008–2010
National Council of Teachers of Mathematics

Preface

Although statistics has many distinct characteristics that distinguish it from the other mathematical sciences, it has a place under the banner of school mathematics. The introduction to this book reflects this convention and refers to *mathematics*, often in the sense of school mathematics. Key differences between statistics and mathematics are evident in the chapters that follow, and the enriched understanding of statistics encouraged by this book should help teachers make the distinction between the two disciplines for themselves and with their students.

From prekindergarten through grade 12, the school mathematics curriculum includes important topics that are pivotal in students' development. Students who understand these ideas cross smoothly into new mathematical terrain and continue moving forward with assurance.

However, many of these topics have traditionally been challenging to teach as well as learn, and they often prove to be barriers rather than gateways to students' progress. Students who fail to get a solid grounding in them frequently lose momentum and struggle in subsequent work in mathematics and related disciplines.

The Essential Understanding Series identifies such topics at all levels. Teachers who engage students in these topics play critical roles in students' mathematical achievement. Each volume in the series invites teachers who aim to be not just proficient but outstanding in the classroom—teachers like you—to enrich their understanding of one or more of these topics to ensure students' continued development in mathematics.

How much do you need to know?

To teach these challenging topics effectively, you must draw on a mathematical understanding that is both broad and deep. The challenge is to know considerably more about the topic than you expect your students to know and learn.

Why does your knowledge need to be so extensive? Why must it go above and beyond what you need to teach and your students need to learn? The answer to this question has many parts.

To plan successful learning experiences, you need to understand different models and representations and, in some cases, emerging technologies as you evaluate curriculum materials and create lessons. As you choose and implement learning tasks, you need to know what to emphasize and why those ideas are mathematically important.

While engaging your students in lessons, you must anticipate their perplexities, help them avoid known pitfalls, and recognize and dispel misconceptions. You need to capitalize on unexpected classroom opportunities to make connections among mathematical ideas. If assessment shows that students have not understood the material adequately, you need to know how to address weaknesses that you have identified in their understanding. Your understanding must be sufficiently versatile to allow you to represent the mathematics in different ways to students who don't understand it the first time.

In addition, you need to know where the topic fits in the full span of the mathematics curriculum. You must understand where your students are coming from in their thinking and where they are heading mathematically in the months and years to come.

Accomplishing these tasks in mathematically sound ways is a tall order. A rich understanding of the mathematics supports the varied work of teaching as you guide your students and keep their learning on track.

How can the Essential Understanding Series help?

The Essential Understanding books offer you an opportunity to delve into the mathematics that you teach and reinforce your content knowledge. They do not include materials for you to use directly with your students, nor do they discuss classroom management, teaching styles, or assessment techniques. Instead, these books focus squarely on issues of mathematical content—the ideas and understanding that you must bring to your preparation, in-class instruction, one-on-one interactions with students, and assessment.

How do the authors approach the topics?

For each topic, the authors identify "big ideas" and "essential understandings." The big ideas are mathematical statements of overarching concepts that are central to a mathematical topic and link numerous smaller mathematical ideas into coherent wholes. The books call the smaller, more concrete ideas that are associated with each big idea *essential understandings*. They capture aspects of the corresponding big idea and provide evidence of its richness.

The big ideas have tremendous value in mathematics. You can gain an appreciation of the power and worth of these densely packed statements through persistent work with the interrelated essential understandings. Grasping these multiple smaller concepts and through them gaining access to the big ideas can greatly increase your intellectual assets and classroom possibilities.

In your work with mathematical ideas in your role as a teacher, you have probably observed that the essential understandings are often at the heart of the understanding that you need for presenting one of these challenging topics to students. Knowing these ideas very well is critical because they are the mathematical pieces that connect to form each big idea.

Big ideas and essential understandings are identified by icons in the books.

marks a big idea, and

marks an essential understanding.

Preface

How are the books organized?

Every book in the Essential Understanding Series has the same structure:

- The introduction gives an overview, explaining the reasons for the selection of the particular topic and highlighting some of the differences between what teachers and students need to know about it.

- Chapter 1 is the heart of the book, identifying and examining the big ideas and related essential understandings.

- Chapter 2 reconsiders the ideas discussed in chapter 1 in light of their connections with mathematical ideas within the grade band and with other mathematics that the students have encountered earlier or will encounter later in their study of mathematics.

- Chapter 3 wraps up the discussion by considering the challenges that students often face in grasping the necessary concepts related to the topic under discussion. It analyzes the development of their thinking and offers guidance for presenting ideas to them and assessing their understanding.

The discussion of big ideas and essential understandings in chapter 1 is interspersed with questions labeled "Reflect." It is important to pause in your reading to think about these on your own or discuss them with your colleagues. By engaging with the material in this way, you can make the experience of reading the book participatory, interactive, and dynamic.

marks a "Reflect" question that appears on a different page.

Reflect questions can also serve as topics of conversation among local groups of teachers or teachers connected electronically in school districts or even between states. Thus, the Reflect items can extend the possibilities for using the books as tools for formal or informal experiences for in-service and preservice teachers, individually or in groups, in or beyond college or university classes.

A new perspective

The Essential Understanding Series thus is intended to support you in gaining a deep and broad understanding of mathematics that can benefit your students in many ways. Considering connections between the mathematics under discussion and other mathematics that students encounter earlier and later in the curriculum gives the books unusual depth as well as insight into vertical articulation in school mathematics.

The series appears against the backdrop of *Principles and Standards for School Mathematics* (NCTM 2000), *Curriculum Focal Points for Prekindergarten through Grade 8 Mathematics: A Quest for Coherence* (NCTM 2006), *Focus in High School Mathematics: Reasoning and Sense Making* (NCTM 2009), and the Navigations

Series (NCTM 2001–2009). The new books play an important role, supporting the work of these publications by offering content-based professional development.

The other publications, in turn, can flesh out and enrich the new books. After reading this book, for example, you might select hands-on, Standards-based activities from the Navigations books for your students to use to gain insights into the topics that the Essential Understanding books discuss. If you are teaching students in prekindergarten through grade 8, you might apply your deeper understanding as you present material related to the three focal points that *Curriculum Focal Points* identifies for instruction at your students' level. Or if you are teaching students in grades 9–12, you might use your understanding to enrich the ways in which you can engage students in mathematical reasoning and sense making as presented in *Focus in High School Mathematics*.

An enriched understanding can give you a fresh perspective and infuse new energy into your teaching. We hope that the understanding that you acquire from reading the book will support your efforts as you help your students grasp the ideas that will ensure their mathematical success.

We wish to thank the reviewers of this volume who provided valuable suggestions about the format and content of the book. Henry Kranendonk, Susan A. Peters, Toni Smith, and Mike Shaughnessy gave freely of their time and expertise to help the authors work through the difficult task of keeping the discussion rigorous and accessible. Everyone contributed to the authors' efforts to organize an enormous amount to material into a few big ideas and an accompanying discussion that would be relevant and helpful for teachers working with high school students.

Introduction

This book focuses on ideas about statistics. These are ideas that you need to understand thoroughly and be able to use flexibly to be highly effective in your teaching of mathematics in grades 9–12. The book discusses many statistical ideas that are common in high school curricula, and it assumes that you have had a variety of experiences with statistics that have motivated you to delve into—and move beyond—the statistics that you expect your students to learn.

The book is designed to engage you with these ideas, helping you to develop an understanding that will guide you in planning and implementing lessons and assessing your students' learning in ways that reflect the full complexity of statistical concepts. A deep, rich understanding of these concepts will enable you to communicate their influence and scope to your students, showing them how these ideas permeate the statistics that they have encountered—and will continue to encounter—throughout their experiences with statistics.

The understanding of statistics that you gain from this focused study thus supports the vision of *Principles and Standards for School Mathematics* (NCTM 2000): "Imagine a classroom, a school, or a school district where all students have access to high-quality, engaging mathematics instruction" (p. 3). This vision depends on classroom teachers who "are continually growing as professionals" (p. 3) and routinely engage their students in meaningful experiences that help them learn statistics with understanding.

Why Statistics?

Like the topics of all the volumes in NCTM's Essential Understanding Series, statistical procedures and concepts compose a major area of school mathematics that is crucial for students to learn but challenging for teachers to teach. Students in grades 9–12 need to understand statistical concepts well if they are to succeed in these grades and in their subsequent experiences with statistics. Learners often struggle with ideas about statistics. For example, they are puzzled about what conclusions can be drawn from a sample when the sample is only a small fraction of a large population. Teachers of statistics in grades 9–12 understand the characteristics of good sampling methods and the concepts of bias and precision and are able to use their understanding to help students value the use of samples in statistical inference.

Your work as a teacher of mathematics in these grades calls for a solid understanding of the statistical concepts that you—and your school, your district, and your state curriculum—expect your students to learn. Your work also requires you to know how these ideas relate to other statistical and mathematical ideas that your students will encounter in the lesson at hand, the current school year, and beyond. A rich understanding of statistical concepts guides teachers' decisions in much of their work, such as choosing tasks for a lesson, posing questions, selecting materials, ordering topics and ideas over time, assessing the quality of students' work, and devising ways to challenge and support their thinking.

Understanding Statistical Concepts

Teachers teach statistics because they want others to understand it in ways that will contribute to success and satisfaction in school, work, and life. Helping your students develop a robust and lasting understanding of statistical concepts requires that you understand statistics deeply. But what does this mean?

It is easy to think that understanding statistics means knowing certain processes for analyzing data and mastering relevant vocabulary. For example, you are expected to be able to calculate a standard deviation, determine a standard error, and construct a confidence interval. You know different sampling strategies, such as simple random sampling and stratified random sampling. You are expected to use terms such as *random assignment*, *confidence interval*, and *precision*.

Obviously, vocabulary and processes are not all that you are expected to know about statistics. For example, in your ongoing work with students, you have undoubtedly discovered that you need not only to know common statistical procedures but also to be able to follow strategies that your students create. You need to make subtle distinctions between terms such as *random assignment* and *random selection*, and *sample distribution* and *sampling distribution*.

It is also easy to focus on a very long list of statistical ideas that all teachers of mathematics in grades 9–12 are expected to know and teach about statistics. Curriculum developers often devise and publish such lists. However important the individual items might be, these lists cannot capture the essence of a rich understanding of the topic. Understanding statistics deeply requires you not only to know important statistical ideas but also to recognize how these ideas relate to one another. Your understanding continues to grow with experience and as a result of opportunities to embrace new ideas and find new connections among familiar ones.

Furthermore, your understanding of statistics should transcend the content intended for your students. Some of the differences between what you need to know and what you expect them to learn are easy to point out. For example, you need not only to know how to specify statistical models by using familiar mathematical notation and to quantify goodness of fit, but also to see how these ideas underlie more complex statistical models, such as multiple regression models.

Other differences between the understanding that you need to have and the understanding that you expect your students to acquire are less obvious, but your experiences in the classroom have undoubtedly made you aware of these differences at some level. For example, how many times have you been grateful to have an understanding of statistics that enables you to recognize the merit in a student's unanticipated question or claim? How many other times have you wondered whether you could be missing such an opportunity or failing to use it to full advantage because of a gap in your knowledge?

As you have almost certainly discovered, knowing and being able to engage in the statistical problem-solving process is not enough when you are in the classroom. You also need to be able to identify and justify or refute novel claims. These claims and justifications might draw on ideas or techniques that are beyond the mathematical experiences of your students and current curricular expectations for them. For example, you need to know more complicated mathematical models than students encounter in algebra and geometry and be able to explain how these models are similar to yet different from statistical models that account for both structure and variation.

Big Ideas and Essential Understandings

Thinking about the many particular ideas that are part of a rich understanding of statistics can be an overwhelming task. Articulating all of those statistical ideas and their connections would require many books. To choose which ideas to include in this book, the authors considered a critical question: What is e*ssential* for teachers of mathematics in grades 9–12 to know about statistics to be effective in the classroom? To answer this question, the authors drew on a variety of resources, including personal experiences, the expertise of colleagues in mathematics, mathematics education, statistics, and statistics education, and the reactions of reviewers and professional development providers, as well as ideas from curricular materials and research on mathematics learning and teaching.

As a result, the content of this book focuses on essential ideas for high school teachers about statistics. In particular, chapter 1 is organized around five big ideas related to this important area. Each big idea is supported by smaller, more specific ideas, which the book calls *essential understandings*.

Benefits for Teaching, Learning, and Assessing

Understanding statistics can help you implement the Teaching Principle enunciated in *Principles and Standards for School Mathematics*. This Principle sets a high standard for instruction: "Effective mathematics teaching requires understanding what students know and need to learn and then challenging and supporting them to learn it well" (NCTM 2000, p. 16). As in teaching about other critical topics, teaching about statistics requires knowledge that goes "beyond what most teachers experience in standard preservice mathematics courses" (p. 17).

Chapter 1 comes into play at this point, offering an overview of the topic that is intended to be more focused and comprehensive than many discussions that you are likely to have encountered. This chapter enumerates, expands on, and gives examples of the big ideas and essential understandings related to statistics, with the goal of supplementing or reinforcing your understanding. Thus, chapter 1 aims to prepare you to implement the Teaching Principle fully as you provide the support and challenge that your students need for robust learning about statistics.

Consolidating your understanding in this way also prepares you to implement the Learning Principle outlined in *Principles and Standards*: "Students must learn mathematics with understanding, actively building new knowledge from experience and prior knowledge" (NCTM 2000, p. 20). To support your efforts to help your students learn about the concepts in this way, chapter 2 builds on the understanding that chapter 1 communicates by pointing out specific ways in which the big ideas and essential understandings connect with mathematics that students typically encounter earlier or later in school. This chapter supports the Learning Principle by emphasizing longitudinal connections in students' learning about statistics. For example, as their statistical experiences expand, students gain a richer understanding of variability. They move from understanding variability in the case of one variable to considering variability in the case of bivariate models. They can understand the idea of data varying about a line or curve by building on a firm understanding of data varying about the mean of a univariate data set. Understanding variability in general will help students as they move to more complicated statistical models and explore variability that involves multiple variables.

The understanding that chapters 1 and 2 convey can strengthen another critical area of teaching. Chapter 3 addresses this area, building on the first two chapters to show how an understanding of statistics can help you select and develop appropriate tasks, techniques, and tools for assessing your students' understanding of statistics. An ownership of the big ideas and essential understandings related to statistics, reinforced by an understanding of students' past and future experiences with related ideas, can help you ensure that assessment in your classroom supports the learning of foundational statistical concepts.

Such assessment satisfies the first requirement of the Assessment Principle set out in *Principles and Standards*: "Assessment should support the learning of important mathematics and furnish useful information to both teachers and students" (NCTM 2000, p. 22). An understanding of statistics can also help you satisfy the second requirement of the Assessment Principle, by enabling you to develop assessment tasks that give you specific information about what your students are thinking and what they understand. For example, in developing questions for class discussion or for an exam, you can move beyond items that require students only to display their computational skills. You are able to use situations and questions that include real-world contexts and that ask students for interpretations of numerical results and demand explanations that require conceptual understanding.

Ready to Begin

This introduction has painted the background, preparing you for the big ideas and associated essential understandings related to statistics that you will encounter and explore in chapter 1. Reading the chapters in the order in which they appear can be a very useful way to approach the book. Read chapter 1 in more than one sitting, allowing time for reflection. Take time also to use a graphing calculator or other computational tools as you consider items that recommend technology use. Absorb the ideas—both big ideas and essential understandings—related to statistics. Appreciate the connections among these ideas. Carry your newfound or reinforced understanding to chapter 2, which guides you in seeing how the ideas related to these operations are connected to the statistics that your students have encountered earlier or will encounter later in school. Then read about teaching, learning, and assessment in chapter 3.

Alternatively, you may want to take a look at chapter 3 before engaging with the statistical ideas in chapters 1 and 2. Having the challenges of teaching, learning, and assessment issues clearly in mind, along with possible approaches to them, can give you a different perspective on the material in the earlier chapters. No matter how you read the book, let it serve as a tool to expand your understanding, application, and enjoyment of statistics.

Two appendixes supplement the text. The first, a glossary of common statistical terms, is intended for quick reference and clarification as readers move through the text. Terms that are included in the glossary are shown in green in the text. The second appendix, a short list of resources for teachers, is designed to point to information that can extend, reinforce, and enrich readers' experience of the book. Both appendixes are also available online at www.nctm.org/more4u.

Chapter 1

Statistics: The Big Ideas and Essential Understandings

Statistics has been a recommended part of the high school mathematics curriculum for many years. *Curriculum and Evaluation Standards* (NCTM 1989) addressed statistics in Standard 10 of the Curriculum Standards for grades 9–12. *Principles and Standards for School Mathematics* (NCTM 2000) reaffirmed statistics as part of the high school mathematics curriculum with the inclusion of the Data Analysis and Probability Standard.

However, widespread implementation of these statistics standards has been elusive, and many schools and districts that have included statistics and data analysis in the curriculum have incorporated them in an ad hoc fashion. As a consequence, students and teachers often see statistics as a loose collection of graphical and numerical methods with no underlying, unifying theory. Although students exposed to statistics in this manner might be able to construct graphical displays and compute numerical summaries of data, they often develop only a superficial understanding of important statistical concepts and fail to build the kind of statistical reasoning skills that have become essential to making informed decisions in today's quantitative world.

See *Developing Essential Understanding of Statistics for Teaching Mathematics in Grades 6–8* (Kader and Jacobbe 2013) for an extended discussion of statistics as a problem-solving process that involves graphical and numerical methods.

The landscape of statistics in the high school mathematics curriculum is changing in a profound way. With the adoption of the Common Core State Standards for Mathematics (CCSSM; Common Core State Standards Initiative 2010), statistics will take a more prominent place in the high school curriculum. Successful implementation of the Common Core State Standards will require that the high school mathematics curriculum include statistics content that goes beyond the mechanical and computational aspects of descriptive statistical methods to focus on the conceptual understanding necessary for the development of sound statistical reasoning.

In this book, the five big ideas that we have identified focus on concepts that illustrate the use of probabilistic reasoning in statistical inference. We have chosen to focus on concepts rather than on the mechanical and computational aspects of specific inferential methods for two reasons. First, in a book as brief as this one, it is not possible to offer a textbook-style treatment of introductory statistics. Second, we believe that many high school mathematics teachers are comfortable with the mechanical and computational aspects of data analysis but may be somewhat less comfortable with the concepts that provide a unifying structure supporting the use of probabilistic reasoning in statistical inference. Our hope is that the book will allow teachers who are experienced in teaching algebra 1, geometry, and algebra 2 to successfully meet the challenges that they face in integrating a statistics component into these courses as they implement the Common Core State Standards.

> Terms shown in green are defined in the glossary that appears in appendix 1, which is also available at www.nctm.org/more4u.

Big Ideas and Associated Understandings

This book focuses on five big ideas, which we summarize briefly below before listing them with all of their associated essential understandings. After this quick overview of the big ideas and the complete list that follows it, we will devote the remainder of chapter 1 to developing each big idea and essential understanding in detail.

The big ideas in brief

The concept of a mathematical model is ubiquitous in the modern mathematics curriculum. Big Idea 1 draws a distinction between mathematical models and statistical models. For example, in looking at weights of U.S. pennies, we might make a mathematical model that specifies the weight of a penny as 2.5 grams (which is what the U.S. Mint claims as the weight). But the weights of individual pennies will differ, if only slightly, from this value, and a statistical model would extend the mathematical model by incorporating the variability about this central value of 2.5 grams.

A mathematical model might also specify or describe a relationship (for example, a linear relationship) between variables in a bivariate situation. Statistical models build on such mathematical models by explicitly including descriptions of random variation. Statistical methods can then be used to acknowledge and to quantify this variability, allowing us to evaluate the usefulness of a particular model in describing data in terms of structure plus variability.

Understanding variability is necessary for collecting, describing, analyzing, and drawing conclusions from data in a sensible way. Thinking about data in terms of distributions and distinguishing among the different ways to use distributions (to describe the values in a population, the values in a sample, or the values of a *statistic* for different possible samples) are key to understanding statistical inference. The idea that distributions describe variability is the fundamental concept in Big Idea 2.

The reasoning involved in carrying out a *hypothesis test* is simple and intuitive, but it is easy to lose sight of this simple logic in the development of a formal (and often mechanical) process for carrying out such a test. Big Idea 3 focuses on the way in which these tests are used to determine whether a particular outcome might have happened by chance.

Statistical methods involve using available, but usually incomplete, information to draw conclusions about a *population* or about the effect of experimental conditions on some response. Because such conclusions are based on incomplete information, they have an associated risk of error, and this potential needs to be acknowledged and quantified. Characteristics of the data collection method (such as *random selection* and *random assignment*) make this calculation possible. In addition, both the type of analysis that is appropriate and the conclusions that can be drawn depend on the type of data and the way in which they are collected. Data collection methods and their implications are the focus of Big Idea 4.

Understanding that a relatively small sample from a population can be used to make accurate estimates of the characteristics of the entire population is critical to understanding sampling and inference. The size of the sample affects the *precision* with which estimates can be made or claims tested; when the sampling plan is properly designed, a larger sample will always produce more *precise* results than a smaller sample. The fact that the size of the population is not an important factor in determining the *accuracy* of estimates runs counter to people's intuition that a larger population requires a larger sample. The key to understanding how sampling and inference are related depends on the sampling method, the sample size, and the interplay between them. The roles of sampling method, precision, and possible bias in the evaluation of an estimator are the focus of Big Idea 5.

The complete list of ideas and understandings

The list of the big ideas and their associated essential understandings that appears below offers a preview of the detailed discussion of each big idea to follow. Additionally, we present the list here for your convenience in referring back to the big ideas and related understandings as you move through the book. Read through them now, but do not think that you must absorb them fully at this point; we will discuss each one in turn in detail.

Big Idea 1. Data consist of structure and variability.

Essential Understanding 1a. Mathematical models describe structure.

Essential Understanding 1b. Statistical models extend mathematical models by describing variability around the structure.

Essential Understanding 1c. Statistical models are evaluated by how well they describe data and whether they are useful.

Big Idea 2. Distributions describe variability.

Essential Understanding 2a. A population distribution describes variability in the values that make up a population.

Essential Understanding 2b. The population distribution is often unknown but can be approximated by a sample distribution.

Essential Understanding 2c. The sampling distribution of a sample statistic describes how the value of the statistic varies from sample to sample.

Essential Understanding 2d. Simulation can be used to approximate sampling distributions.

Big Idea 3. Hypothesis tests answer the question, "Do I think that this could have happened by chance?"

Essential Understanding 3a. A hypothesis test involves choosing between two competing hypotheses—the null hypothesis and the alternative hypothesis.

Essential Understanding 3b. The alternative hypothesis is determined by the statistical question of interest.

Essential Understanding 3c. The null hypothesis is rejected in favor of the alternative hypothesis if the sample data provide convincing evidence against the null hypothesis.

Essential Understanding 3d. The *p*-value measures surprise.

Essential Understanding 3e. Hypothesis tests do not always lead to a correct decision.

Big Idea 4. The way in which data are collected matters.

Essential Understanding 4a. Observational studies, including surveys, provide information about the characteristics of a population or sample, whereas controlled experiments provide information about treatment effects.

Essential Understanding 4b. Random assignment in an experiment permits drawing causal conclusions about treatment effects and quantifying the uncertainty associated with these conclusions.

Essential Understanding 4c. Random selection tends to produce samples that are representative of the population, permitting generalization from the sample to the larger population and also allowing the uncertainty in estimates to be quantified.

Essential Understanding 4d. Random selection and random assignment are different things, and the type and scope of conclusions that can be drawn from data depend on the role of random selection and random assignment in the study design.

Big Idea 5. Evaluating an estimator involves considering bias, precision, and the sampling method.

Essential Understanding 5a. Estimators are evaluated on the basis of their performance in repeated sampling.

Essential Understanding 5b. Some estimators are biased.

Essential Understanding 5c. The standard error describes the precision of an estimator.

Essential Understanding 5d. Confidence intervals are estimators that convey information about precision.

Essential Understanding 5e. The precision of estimators depends both on the way in which the sample was selected and on the size of the sample.

Essential Understanding 5f. If the sampling method is good, a larger sample is always more useful than a smaller sample.

Essential Understanding 5g. A small sample selected by using a good method can yield better results than a large sample selected by using a poor method.

Essential Understanding 5h. The size of the sample relative to the population size is not an important factor in determining the accuracy of estimates.

The Nature of Statistical Models: Big Idea 1

Big Idea 1. *Data consist of structure and variability.*

Mathematical models are a part of mathematics education because they help students see the relationship between mathematics and the physical world and because they motivate students while deepening their understanding of mathematics. Statistical models serve the same purpose, and combine mathematical models with descriptions of variability. Statistical models for both univariate (single-variable) and bivariate (two-variable) data take the form

$$\text{data} = \text{structure} + \text{variability}.$$

These models include variability in order to provide a more thorough description of the behavior of the data under study. For example, univariate data values can be modeled as varying about a single numeric value such as the mean, often denoted by μ. The single value is said to provide the *structure*, and a statistical model incorporates *variability* in addition to the structure. A statistical model for each data value (y) might be $y = \mu + \varepsilon$, where ε represents a deviation from the mean and therefore describes variability. Bivariate data values can be modeled as varying about a line or curve. For example, a statistical model for a bivariate data value (x, y) might be $y = f(x) + \varepsilon$, where $f(x)$ is a linear function that describes the structure and ε represents a deviation from the function $f(x)$.

One of the early uses of a bivariate statistical model occurred in 1903 when Karl Pearson fitted a model that described the relationship between an adult son's height and his father's height as

$$\text{son's height} = 35 + 0.5 \times (\text{father's height}) + \varepsilon$$

(Freedman, Pisani, and Purves 2007). The structure specifies a linear relationship in which a son's height can be estimated by multiplying the father's height by 0.5 and adding 35 inches. Individual sons, however, might differ from this "deterministic" specification by being somewhat taller or shorter than this amount. The model tells us that an individual son's height, y_i, may differ by an amount ε_i from this linear structure (by convention, the subscript i indicates a particular individual).

Statistical models are evaluated in a variety of ways. We assess models according to their usefulness and effectiveness in describing the data, and we perform this analysis by quantifying variability. If there is relatively little variability about the structural component, then the model is generally, but not always, considered more useful than if there is more variability about the structure. In addition, with bivariate data, we can use particular methods to evaluate how well the structural component captures the overall pattern of the relationship. Three essential understandings regarding the nature and quality of statistical models and mathematical models support Big Idea 1.

Mathematical models describe structure

Essential Understanding 1a. Mathematical models describe structure.

Consider a small population consisting of the heights of 43 statistics students, with all the heights reported to the nearest inch. In the dot plot shown in figure 1.1, each value in the population is represented by a single dot, positioned in the appropriate location along a horizontal measurement scale.

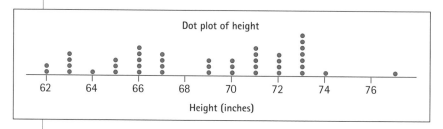

Fig. 1.1. A dot plot of the heights of statistics students

One commonly used measure of the center of data such as these is the mean, or arithmetic average—in this case, the mean height. For this population, the mean height is $\mu = 68.7$ inches. The mean describes a structural aspect of the population—namely, the center of the data. In the case of univariate data, the term *structure* refers to a single numeric value that describes a central location for the overall population. This structural value provides some information about the overall population, but it doesn't give the whole picture. Reflect 1.1 explores the fact that the structure is only part of the picture.

Reflect 1.1

Create three different populations with ten values and the same mean, but with different amounts of variability. If you created dot plots of your three populations, how would the dot plots differ? How would they be similar?

Numerous populations could satisfy the requirements specified in Reflect 1.1. Suppose we want the mean to be 5. Figure 1.2 presents three such hypothetical populations.

Although these three populations have the same mean, the actual values that make up the populations are quite different. The dot plots in figure 1.3 nicely illustrate the differences in the populations. Reflect 1.2 highlights the need for explicitly modeling variability.

The Big Ideas and Essential Understandings

Population 1	5	5	5	5	5	5	5	5	5	5
Population 2	0	0	0	0	0	10	10	10	10	10
Population 3	3	4	4	5	5	5	5	6	6	7

Fig. 1.2. Three populations with the same mean and different amounts of variability

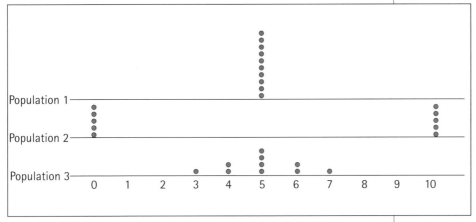

Fig. 1.3. Dot plots of the populations from figure 1.2

Reflect 1.2

On the basis of the dot plots in figure 1.3, which population has the most variability, and which has the least? How do you know?

The mean tells only part of the story. Some description or numeric measure of variability is needed to indicate that population 1 is the least variable and population 2 is the most variable. Knowing only the structural component $\mu = 68.7$ inches for the population of student heights tells us nothing about the variability in heights. A statistical model includes this variability by modeling heights as $y_i = 68.7 + \varepsilon_i$, where ε_i represents the deviation from the mean height for the *i*th data value. For example, if the smallest height in the population is 62 inches, the ε for that height is $62 - 68.7$, or -6.7, inches. Suppose that another height in the population is 73 inches, which is greater than the mean. In this case, ε for that height is 4.3 inches. Heights closer to the mean have ε values that are smaller in magnitude; those farther from the mean have ε values that are larger in magnitude.

Bivariate, or two-variable, populations require functions of the form $y = f(x)$ to model the structure, which is the overall **trend** or relationship between two quantitative variables. The independent variable (sometimes also called the *predictor* or *explanatory* variable) and the dependent variable (sometimes also called the *response variable*) are often denoted by x and y, respectively, but other symbols can be used. For example, Boyle's law

$$P = \frac{k}{V}$$

models the inverse relationship between the pressure (P) and the volume (V) of an ideal gas held at constant temperature (k is a constant of proportionality). Another example is Hooke's law, which models the direct relationship between the extension (x) of a spring and the weight (F) of an object hung on it by the formula $F = kx$, where k is a constant of proportionality called the *spring constant*. These are fairly typical mathematical models encountered in high school math or physics classes, in which one quantity—in these examples, P or F—is completely determined by another quantity—here, V or x. The mathematical models in these examples perform the essential job of describing the relationship that theory tells us exists between two variables. These models, however, do not take into account the variability that students are likely to see in their physics lab when they collect their own data. That variability arises from measurement error, from variations in the equipment, or variations in the settings in which the measurements are taken.

One common structural relationship between x and y is linear, with the mathematical model taking the form $f(x) = a + bx$, where a and b are the y-intercept and slope of the line, respectively. Mathematicians typically write linear equations in the form $f(x) = mx + b$; statisticians, by contrast, tend to use $f(x) = a + bx$ when writing the equation of a line. If a curve more accurately describes the relationship between the variables, then a nonlinear function may be used to model the structure. Some examples of nonlinear functions are exponential, power, and logarithmic functions.

Consider, for example, the relationship between the height and arm span of humans. The ancient Roman architect Marcus Vitruvius postulated that height and arm span in humans are approximately equal. Reflect 1.3 explores the mathematical implications of Vitruvius's conjectured relationship.

Reflect 1.3

If Vitruvius's conjecture that the height and the arm span in humans are approximately equal is indeed correct, what is the equation of the line that models the structural part of that relationship?

If height and arm span are approximately equal, then the function that models the structural part of the relationship is linear, with a slope of 1, and a y-intercept of 0. Specifically, the equation of the line is $y = 0 + 1x$, where y represents the height and x represents the arm span. This function indicates that all the data values lie along the line $y = x$, with no variability. However, in reality, not all people have height exactly equal to arm span. As a result, there is variability about the line $y = x$, with some points lying above the line, and others lying below it. In addition, some of the points may be relatively far from the line, and others may be relatively close to it. Once again, we can apply the data = structure + variability approach. The line provides the structural component, but the actual data values vary above and below that line. The variability can be due to the approximate nature of the model because not all humans have arm spans equal to their heights, but variability can also be due in part to measurement error associated with measuring lengths.

For a discussion of underlying ideas of covariability and associations between two variables, see *Developing Essential Understanding of Statistics for Teaching Mathematics in Grades 6–8* (Kader and Jacobbe 2013).

Statistical models include variation

Essential Understanding 1b. *Statistical models extend mathematical models by describing variability around the structure.*

The structures of univariate and bivariate data sets are modeled mathematically by using either measures of location, such as the mean, or functions that describe the underlying relationship between two variables. The structural nature of mathematical models is somewhat limiting in that these models tell only part of the story. Statistical models extend the mathematical model by including a variability component. Statistical measures of variability, such as the **standard deviation**, give an indication of how much, on average, data values deviate from the structural part of the model. In the univariate case, variability is usually measured in terms of deviations from the mean. In the case of bivariate data, the variability is usually measured in terms of an average deviation from a line or curve. Variability is a difficult concept for many students (delMas and Liu 2005). Reflect 1.4 provides an entry point into this challenging concept.

> ### Reflect 1.4
>
> For univariate data, what would a mathematical measure that describes a typical deviation from the mean look like?

Recall the three populations in figure 1.2 with the same mean but different amounts of variability. In population 1, all the values are identical, with each value equal to the mean of 5. In this case, there is no deviation from the mean. However, in populations 2 and 3, the data values vary about the mean. How can we measure this variability? A frequently used measure of variability is the **population standard deviation**, often denoted by σ. The standard deviation is a measure of a typical distance from the mean. Note that there are other ways to measure variability, but this is one of the most commonly used measures. In our contrived populations, the standard deviation of population 1 is 0, the standard deviation of population 2 is 5, and the standard deviation of population 3 is 1.10. Note that the population with the largest standard deviation has values that are, on average, farthest from the mean.

We can now analyze the data on student height (see fig. 1.1) in more detail. Recall that the mean height of the 43 students is $\mu = 68.7$ inches. Suppose that we also know that the standard deviation is $\sigma = 3.84$ inches. What additional information does the standard deviation provide? It indicates that, *on average*, the heights vary about the mean by about 3.84 inches. The dot plot in figure 1.4 shows the distribution of heights.

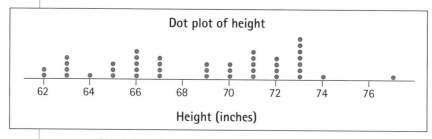

Fig. 1.4. Dot plot of the heights of statistics students

Referring again to the relationship postulated by Vitruvius between arm span and height, we note that if the relationship between the variables were perfectly linear, with arm span and height exactly equal to each other for each person, the points on the scatterplot would lie along line $y = x$. In reality, the points scatter about the line. The plot in figure 1.5 illustrates arm span–height data, in centimeters, collected for a population consisting of 40 algebra students. Note that the points do follow a linear trend, illustrated by the line, and that there is scatter about the line.

The amount of scatter is quantified by the standard deviation of the vertical deviations between the points and the line. Most introductory statistics textbooks explain the concept of fitting lines to data; readers are encouraged to consult one of those books for more detailed information.

Good models give useful descriptions of structure

Essential Understanding 1c. *Statistical models are evaluated by how well they describe data and whether they are useful.*

When mathematical models are used to describe physical phenomena, they are, at best, only approximations of reality. They are generally based on simplifying assumptions that have the benefit

The Big Ideas and Essential Understandings

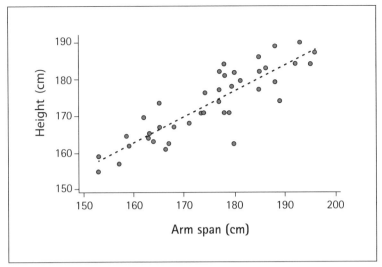

Fig. 1.5. Scatterplot of height and arm span data

of producing a model that is relatively easy to understand, explain, and use. For example, the relationship between the period and length of a pendulum is often studied in mathematics and physics classes. That relationship states that the period of a pendulum—the time for the pendulum to make one complete oscillation—is directly proportional to the square root of the length of the pendulum. The theoretical relationship is

$$T = 2\pi\sqrt{\frac{L}{g}},$$

where T is the period, g is the acceleration due to gravity (approximately 9.81 m/s² or 32 ft/s²), and L is the length of the pendulum. This theoretical relationship models the structure of the relationship between period and length. It is important to note, however, that this theoretical model is for a point pendulum supported by a massless, inextensible cord, a situation that is not realistic. In addition, this model applies only when the angular displacement of the pendulum is no more than approximately 10 to 15 degrees. When students collect data to investigate the relationship between the period and the length of a pendulum by using real (non-point) pendulum bobs and cords with both mass and the potential to stretch, they will observe a scattering of data points about the curve for the mathematical model. There may be other reasons for this scattering, a possibility that Reflect 1.5 explores.

Reflect 1.5

In addition to the unrealistic assumptions required to derive the theoretical model, what other factors might contribute to deviations from what theory predicts?

Other deviations from the theoretical results could be due to buoyancy of the pendulum bob in air, air resistance as the bob swings through the air, and inaccuracies in measurements of the length and period of the pendulum. Statistical models, which are also approximations of reality, extend mathematical models and allow for the incorporation of variability around the structure. The statistical model for the pendulum is

$$T = 2\pi\sqrt{\frac{L}{g}} + \varepsilon.$$

This model also approximates reality. How, then, should we evaluate whether or not it is a good model? Reflect 1.6 invites consideration of this question.

Reflect 1.6

What should be taken into consideration when evaluating a statistical model?

Two factors that should be considered in evaluating statistical models go hand in hand. The first is how well the structural part of the model describes the data. For example, it is possible to use a line to model data that follows a curved pattern, but a curve would provide a much better description of the structure. The second is how useful the model is. Even if the structural part of the model provides a reasonable description of the data, there could still be so much variability around the structure that the model isn't useful. Consider, for example, figure 1.6, in which a scatterplot displays bivariate data on period and length for a pendulum, and a line is overlaid to model the data. Reflect 1.7 illustrates the need for choosing models that provide the appropriate structure.

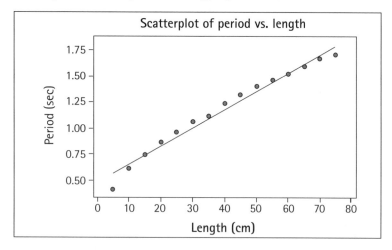

Fig. 1.6. A scatterplot of pendulum data, showing period vs. length, with a line overlaid

The Big Ideas and Essential Understandings

> **Reflect 1.7**
>
> Figure 1.6 illustrates the use of a line to model the relationship between pendulum period and pendulum length. Why is the line not an appropriate model for these data?

Notice that the line is an inappropriate model for the structure, although this is not completely obvious at first glance. The line does indeed capture the fact that longer pendulums tend to have longer periods, but careful inspection of the plot indicates some curvature in the pattern of the data points. The line overestimates the period for small values of length, underestimates the period for moderate values of length, and again overestimates the period for large values of length. Such systematic departures from the model indicate that the line does not accurately capture the structure of the data. The variability component of a statistical model is not meant to describe systematic deviations from the structure—it is the structural component of the model that captures systematic patterns. The variability component of a statistical model describes only nonsystematic (random) deviations from the structure.

Assuming that the data do not deviate from the mathematical model in a systematic fashion, a good model will ideally have small variability with respect to the structure. Small variability translates into an improved ability to make predictions about future observations. For example, figure 1.7 shows two different models that are potentially useful for predicting something called the "Academic Performance Index," or API, for a California high school (http://www.cde.ca.gov/ta/ac/ap/apidatafiles.asp). The first model describes the relationship between a school's API and the percentage of its students' parents who graduated from college. The second model describes the relationship between a school's API and the average education level of its students' parents. (The California Department of Education rescales years of education to a value from 1 to 5 and calls the resulting value the "education level.") Reflect 1.8 asks you to consider which model would be more useful if you had to predict a school's API score on the basis of only one characteristic of that school's student population.

> **Reflect 1.8**
>
> Two models for California high schools' "Academic Performance Index" (API) scores are shown in figure 1.7. Both are good models with respect to describing the structure of the data. Which model do you think would be more useful if you were asked to predict a school's API score on the basis of only one characteristic of that school's student population: the percentage of parents who graduated from college or the parents' average education level?

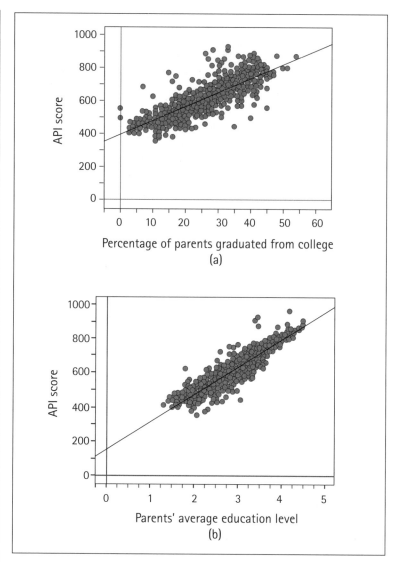

Fig. 1.7. Model (a) indicates that schools' API scores are equal to 395 + 8.5 × (percentage of parents who graduated from college) plus variation. Model (b) indicates that the schools' API scores are equal to 160 + 169 × (parents' average level of education) plus variation.

Note that there is greater variability about the line in the case of the model that uses the percentage of parents who graduated from college, as indicated by the fact that the points are not as tightly clustered about the line. As a result, we would expect predictions of API scores based on the percentage of parents who graduated from college to be less precise than predictions based on the average education level of the parents. The model that uses the average education of the parents would be more useful for predicting a school's API score.

The following statement is often quoted but nevertheless worth repeating here:

> The most that can be expected from any model is that it can supply a useful approximation to reality: All models are wrong; some models are useful. (Box, Hunter, and Hunter 2005, p. 440)

Statistical and mathematical models are, fundamentally, simplifications of reality. The structural component of a statistical model is often based on assumptions that reduce the complexity of the relationship, such as using a measure of center to describe a set of univariate data or using a linear function to describe a set of bivariate data. Because data vary, the structural part of the model by itself is technically incorrect. Models become useful when we can use the structural part to accurately and precisely describe the data, taking variability into account. This can be accomplished more successfully for data with less variability than for data with more variability.

Describing Variability: Big Idea 2

Big Idea 2. *Distributions describe variability.*

We live in a complicated world and often have to make decisions or answer questions that involve some uncertainty. This uncertainty usually arises from variability. In statistics, distributions are used to describe and model variability. Should we purchase an extended warranty for that new laptop? Variability is present in the length of time before a repair is likely to be needed. To make an informed decision about the potential benefit of purchasing the warranty, we need to understand this variability. Would a majority of the students at a school support a ban on cell phones at school? If we could see the distribution of yes and no responses of students at the school, we would know not only the majority preference, but also the degree of consensus. Reflect 1.9 asks you to consider the role that variability plays in choosing between two brands of light bulbs with the lifetimes shown in figures 1.8 and 1.9.

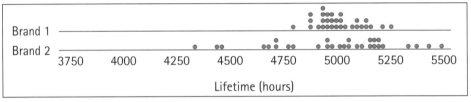

Fig. 1.8. Dot plots of the lifetime (in hours) of 30 light bulbs from two brands

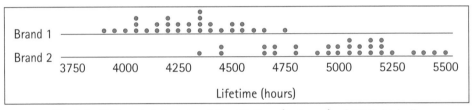

Fig. 1.9. Different dot plots of the lifetime (in hours) of 30 light bulbs

Reflect 1.9

Suppose that you are considering light bulbs from two different brands for the outdoor fixtures at your school. Both brands advertise a mean lifetime of 5000 hours. But not every light bulb functions for exactly 5000 hours before burning out. Thirty bulbs of each brand were tested and the actual lifetime (in hours) for each bulb was recorded and used to make the dot plots in figure 1.8.

a. Which brand would you recommend that your school purchase? What was the basis for your decision?

b. Would your recommendation have been different if the lifetime dot plots had been those shown in figure 1.9? If so, why?

Learning from data and making informed choices on the basis of data depend on understanding and describing variability. Answering questions like those posed in Reflect 1.9 requires that we consider the given data for each brand of light bulbs as a whole—we must look beyond the individual data points to reason about the entire collection, or set, of data points. In the first pair of dot plots, the data values appear to be centered in about the same place, suggesting that a typical lifetime is something around 5000 hours for both brands. The big difference between the two brands is in the variability of the data. Lifetimes are much more consistent—that is, less variable—for bulbs from brand 1. In the second pair of dot plots, the two brands are similar in terms of variability, but the typical lifetime for bulbs from brand 1 is quite a bit shorter than that for bulbs from brand 2.

Big Idea 2, that distributions describe variability, is fundamental to statistics, but it is complex and challenging. Research suggests that making the transition from talking about individual values to considering properties of the entire collection—the distribution—is a key conceptual leap in the development of students' statistical reasoning (for example, see Ben-Zvi and Arcavi [2001] and Hancock, Kaput, and Goldsmith [1992]). Four essential understandings support this "distributional thinking."

Variability in a population

Essential Understanding 2a. A population distribution describes variability in the values that make up a population.

The goal of many statistical studies is to learn about some population of interest. For example, you might be interested in learning about the number of text messages sent in a particular month by students at your school (thus, investigating a **numerical variable**) or about student opinion on a proposal to eliminate student parking on campus (thus, investigating a **categorical variable**). In these examples, the population of interest would be all students enrolled at your school.

Although it is common to think of a population like this as a collection of people, we could also view it as a collection of data values. For example, if we are interested in learning about the number of text messages sent in a particular month, the variable

$$x = \text{number of text messages sent}$$

associates a numerical value with each student. We can then think of the population as being the collection of the x values

$$317 \quad 0 \quad 1214 \quad 798 \quad 0 \quad 0 \quad \ldots$$

To learn about the variability in the values of a variable in a population, we can consider its **population distribution**. Probably the easiest way to do this is to make use of a graphical display like a dot plot or a histogram. A dot plot works well if the population is not too large and is most useful for populations with fewer than, say, 200 observations. For large populations, it is common to use a histogram, which sacrifices some detail about individual data values but still provides a good overall picture of the population distribution.

Suppose that you are interested in the number of text messages sent in the previous month by the 50 students in your two algebra 2 classes. In this case, the population is relatively small and consists of the 50 values of number of text messages sent by these 50 students. Suppose that these values are the following:

0	89	428	0	14	0	178	1021	1800	221	0
0	170	170	405	716	628	116	789	128	4	769
179	582	204	230	25	323	489	568	247	316	407
680	685	0	788	541	363	63	226	527	21	651
579	14	963	549	270	1867					

Summarizing these values in a dot plot gives the picture of the population distribution shown in figure 1.10.

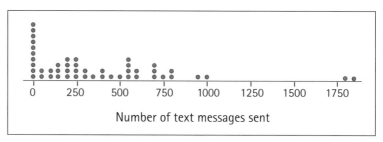

Fig. 1.10. The population distribution of the number of text messages sent by a population of 50 students

What can you learn from this distribution? The first thing that you might notice is a lot of variability in the values that make up this population. This corresponds to student-to-student variability in the number of text messages sent. You might also notice a concentration of values at 0, indicating that quite a few students did not send any text messages. Furthermore, although the mean value for this population is 400, some students sent an unusually large number of text messages compared with most students in the population. Looking at the distribution of values in a way that acknowledges the variability in the population is much more informative than looking only at a single summary measure, such as the population mean value of 400.

The Big Ideas and Essential Understandings

Now suppose that the population of interest is all students at your school. Even if you know the x-value for each and every student, constructing a dot plot of these values would be difficult—dot plots are best for displaying small numerical data sets. For large data sets, you might use a histogram to display the population distribution. For example, figure 1.11 shows a histogram of 2000 values of x = number of text messages sent for a school with 2000 students. Although we can't see all 2000 individual values in the histogram, we can still learn a lot from this population distribution—we can still see the concentration near 0 and note that there are some extreme values on the high side.

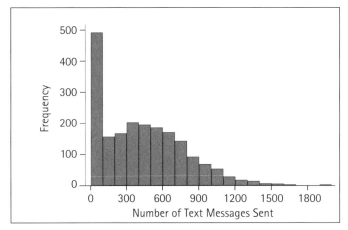

Fig. 1.11. The population distribution of the number of text messages sent for a population of 2000 students

In our discussion of Big Idea 1, we saw that we can model data as

$$\text{data} = \text{structure} + \text{variability}.$$

In the case of univariate data, for an individual data point, the structure might be the population mean, and the variability would then be the deviation from the mean. Such a model is useful if it does a good job of describing the population distribution in terms of important characteristics, such as shape, center, and variability. When describing population distributions, we commonly specify a model (such as a **normal distribution** or an **exponential distribution**) to describe the variability around a central value in the population distribution.

For example, the authors of a paper that appeared in the journal *Ultrasound in Obstetrics and Gynecology* (Lee et al. 2009) used a normal probability distribution with mean of 3500 grams and a standard deviation of 600 grams as a model for the distribution of birth weight for the population of all full-term babies born in the United States. The normal distribution is bell-shaped, symmetric, and centered at its mean. Its standard deviation describes how much the distribution spreads out around the mean. This probability distribution models both the center (3500 grams) and the variability around that central value. The researchers considered this model useful, not because the actual population distribution is *exactly* a normal distribution, but because the population distribution had the same mean and standard deviation as in the model, and the researchers had good reasons for believing that the shape of the population distribution was approximately bell-shaped. Note that a normal distribution would not be a useful model for the distribution of the number of text messages shown in figure 1.11, even if the normal distribution had the same mean and standard deviation, because the shape of the population distribution is far from symmetric.

A population might consist of values of a categorical variable rather than a numerical variable. For example, if we wanted to learn about student opinion on a proposal to ban cell phones on campus, we could consider the variable

$$y = \text{opinion on cell phone ban,}$$

where possible values of y are "support the ban" and "oppose the ban." When the variable of interest is categorical, the population distribution is usually displayed in a bar chart like that in figure 1.12.

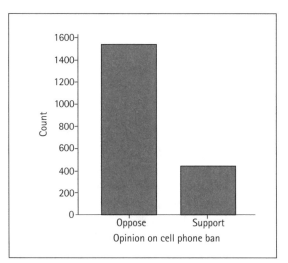

Fig. 1.12. The population distribution of opinions on a proposed cell phone ban for a population of 2000 students

Although we can learn a great deal about a variable by looking at its population distribution, it is usually the case that we do not know the exact population distribution. For example, to construct the population distribution of the number of text messages sent or the opinion on a proposed cell phone ban at a school with 2000 students, we would need to know the values for every individual in the

The Big Ideas and Essential Understandings

population. Unfortunately, this is rarely the case. But it may still be possible to learn about a population by studying a subset of the population.

Approximating population distributions

Essential Understanding 2b. The population distribution is often unknown but can be approximated by a sample distribution.

When the population distribution is unknown and it is not feasible to study the entire population, it is common to focus attention on a subset of the population. The subset of the population selected for study is called a *sample*. Just as with the population, we can regard the sample as a collection of data values and examine its distribution.

For example, suppose that we selected 40 students at random from the population of 2000 students used to construct the population distribution shown in figure 1.11. Suppose that our resulting 40 data values for x = number of text messages sent were the following:

671	827	0	264	0	29	0	297	560	538	59
149	1055	767	388	0	786	1054	300	1079	681	61
752	603	0	672	1067	0	472	135	874	355	42
1008	304	904	0	82	16	37				

Figure 1.13 shows these 40 data values in a dot plot, which displays the distribution of the sample. The *sample distribution* (sometimes also called a *sample data distribution*) describes the variability seen in the 40 values that make up the sample, whereas the population distribution describes the variability in the 2000 data values that make up the entire population. Reflect 1.10 reinforces the distinction between the population distribution and the sample distribution.

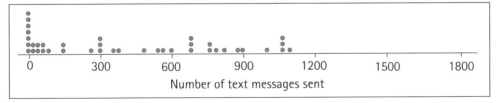

Fig. 1.13. The sample distribution of the number of text messages sent for one sample of size 40

Reflect 1.10

In what ways are the sample distribution shown in figure 1.13 and the population distribution in figure 1.11 similar? In what ways are they different?

Big Idea 4

The way in which data are collected matters.

Big Idea 5

Evaluating an estimator involves considering bias, precision, and the sampling method.

Sample selection is a very important and complex topic that we explore in depth in connection with Big Ideas 4 and 5. Here we briefly note that if a sample is selected *at random* from a population, it is reasonable to expect that the distribution of values in the sample will resemble the population distribution (particularly if the sample size is large). If this is the case, we can use the sample distribution to draw conclusions about the population as a whole.

For example, when the sample distribution shown in figure 1.13 is compared with the population distribution shown in figure 1.11, we see that despite some differences, the sample distribution and the population distribution are quite similar overall. In the sample distribution, we see a concentration of values at 0 and quite a bit of variability in the data values that make up the sample. The mean number of text messages sent for the sample is 422.2, which is reasonably close to the actual value of the mean for the entire population, 420.6.

If we regarded the sample distribution as an approximation of the population distribution and used it to draw conclusions about the population, we would not go too far wrong. The most noticeable difference between the sample distribution and the population distribution is that the sample didn't happen to include any of the unusually large values that occur in the population.

Does the sample distribution always resemble the population distribution? There is no guarantee, but as long as the sample is selected at random from the population and the sample size is not too small, the sample distribution is unlikely to differ from the population distribution in any substantial way. The sample distribution will vary from one sample to another just by chance because different individuals will be chosen when a different sample is selected. But if the sample size is not too small and individuals are selected at random, sample distributions will tend to be similar to one another and also to the population distribution.

Figure 1.14 illustrates this point by showing the population distribution and the sample distributions for six different randomly selected samples of size 40. (For *very* small samples—for example, samples with ten or fewer observations—the shape of the sample distribution may not provide a useful approximation of the shape of the population distribution.)

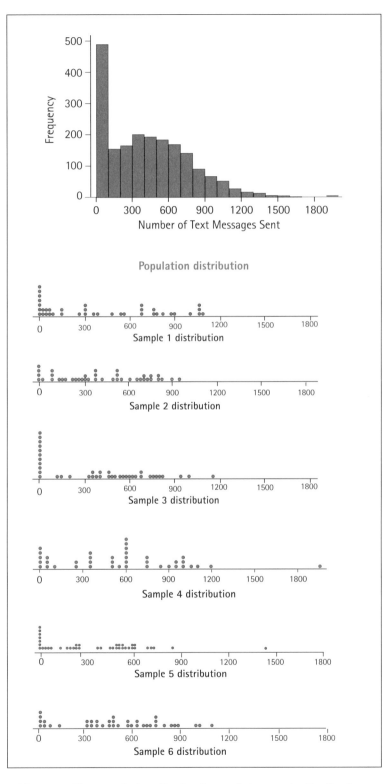

Fig. 1.14. The population distribution and six sample distributions for samples of size 40

Variability in the values of a sample statistic

Essential Understanding 2c. The sampling distribution of a sample statistic describes how the value of the statistic varies from sample to sample.

Often we are interested in learning about a particular population characteristic. Such a characteristic might be the mean value of a numerical variable (like the mean number of text messages sent in a particular month) or the proportion of the individuals in the population who possess some characteristic of interest (like the proportion of individuals who support a cell phone ban). If the population distribution is known, we can compute the exact value of the population characteristic of interest. For example, if we know the population distribution for the number of text messages sent for a population consisting of 2000 students, we can compute the exact value of the population mean. If we do not know the population distribution, we can select a **random sample** from the population and use the sample distribution to approximate the population distribution. We can calculate the mean of the values in the sample, but this sample mean will not usually be exactly equal to the actual population mean. However, if the sample distribution is a reasonable approximation of the population distribution, we can use the value of the sample mean as an approximation of the population mean.

This is the point at which things start to get a bit more complicated. Not only will the sample mean differ from the actual value of the population mean, but different samples will have different sample means. Even when samples are selected at random from the same population, the samples may differ from one another just as a result of the process of **random selection**, so the sample means may also differ. This sample-to-sample variation is called **sampling variability**.

To assess the magnitude of risk involved in using sample data to make statements about the larger population, we need to understand this sampling variability. In particular, we need to understand how much the means calculated from different samples will tend to differ from one another and also how much they will tend to differ from the actual population mean.

To begin to get a sense of this, let's look again at the six different samples of size 40, selected at random from a population consisting of the number of text messages sent for each of 2000 students, shown in figure 1.14. The population mean and the six sample means are as follows:

Population mean	420.6
Sample 1 mean	422.2
Sample 2 mean	418.2
Sample 3 mean	411.7
Sample 4 mean	564.2
Sample 5 mean	367.2
Sample 6 mean	484.7

Notice that two of the sample means are quite close to the value of the population mean, but one in particular—564.2—is quite a bit greater than the population value. Quite a bit of sample-to-sample variability is present in the values of the sample means, and this is something that we will want to consider when we use the mean of a sample to draw conclusions about the population mean. How is sample-to-sample variability affected by the size of the sample? Reflect 1.11 considers this question.

Reflect 1.11

If the sample size for the six samples selected from the population consisting of the 2000 values of number of text messages sent had been 10 instead of 40, do you think you would see more or less sample-to-sample variability? Why do you think so?

The extent of sample-to-sample variability is related to sample size. As you might expect, small samples will tend to differ more from one another and from the population than will large samples, which we would expect to resemble the population distribution more closely.

To get an even better sense of the sample-to-sample behavior of the sample mean for samples of size 40, it would be a good idea to look at more than just six samples. The dot plot in figure 1.15 is a plot of the *sample means* for each of 200 different samples of size 40, selected at random from the population.

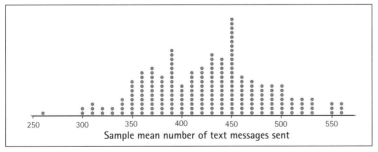

Fig. 1.15. A dot plot of 200 sample means, with each dot representing the mean of one sample of size 40

Carefully examining the dot plot in figure 1.15 tells us a lot about the behavior of the sample mean for samples of size 40 selected at random from this population. Although the 200 sample means tend to center around the actual population mean of 420.6, there is quite a bit of sample-to-sample variability. One sample has a mean around 260, and several samples have means greater than 550. What this tells us is that if we were to take a single sample of size 40, selected at random from this population, we couldn't count on the sample mean being very close to the actual value of the population mean. (Although we acknowledge that this is a subjective judgment, we think that most people would agree that the time needed to send 420 text messages differs substantially from the time needed to send 550, and so in a practical sense, there is quite a bit of variability between these sample means.) Most of the time the sample mean is within 100 of the actual population mean, but, depending on how we plan to use this estimate, that may not be good enough.

The dot plot in figure 1.15 is the start of what is known as a **sampling distribution**. A sampling distribution describes the sample-to-sample variability of a sample statistic, such as the sample mean. What distinguishes the dot plot in figure 1.15 from the actual sampling distribution of the sample mean for samples of size 40, selected at random from this population, is that it is based on the means from only 200 samples. The actual sampling distribution would be based on the means from every possible random sample of size 40. Finding the actual sampling distribution would be a *huge* undertaking—there are more than 9×10^{83} different samples of size 40! Fortunately, there are ways to approximate this sampling distribution.

Approximating sampling distributions

Essential Understanding 2d. *Simulation can be used to approximate sampling distributions.*

Even though we would not want to look at all the different possible samples to find the exact sampling distribution of the sample mean in the example just considered, we are able to get a pretty good idea of what the sampling distribution looks like by considering the distribution of sample means from 200 random samples, as in the dot plot in figure 1.15. Rather than enumerate every possible sample, we constructed the dot plot by *simulating* the sampling process. For this **simulation**, we just needed access to the population to be able to sample from it.

By taking 200 random samples and computing the mean for each of those samples, we are finding just 200 of the sample means that make up the sampling distribution. Because the 200 samples were selected at random, we can think of them as a random sample of all of the different possible random samples. Thus, the dot plot of the 200 sample means provides an approximation of the actual sampling distribution in the same way that a sample distribution provides an approximation of the population distribution. We can get an even better approximation of the sampling distribution by using a larger number of random samples to construct the approximation. For example, figure 1.16 shows a histogram constructed from the sample means from 1000 random samples of size 40.

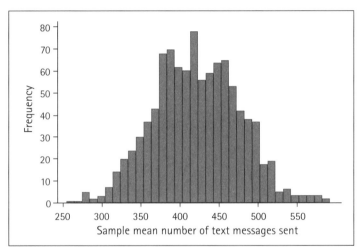

Fig. 1.16. The **approximate sampling distribution** of the sample mean for samples of size 40, drawn from the population shown in figure 1.11

The approximate sampling distribution of the sample mean for samples of size 40 differs from the population distribution in several ways. Figure 1.17 shows both the population distribution and the approximate sampling distribution, overlaid on the same pair of axes. Notice that although the sample means center around the actual population mean of about 421, the shape of the approximate sampling distribution is quite different from that of the population distribution. The approximate sampling distribution is much more symmetric and much less spread out than the population distribution. Reflect 1.12 asks you to think further about the difference between the sampling distribution of a sample mean and the population distribution.

Reflect 1.12

How would a histogram of the sampling distribution of the sample mean compare with the population distribution of number of text messages sent if the sample size were $n = 1$?

The sampling distribution of the sample mean will usually differ from the population distribution in both shape and variability.

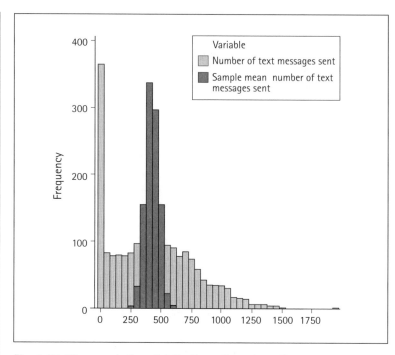

Fig. 1.17. The population distribution of number of text messages sent and the approximate sampling distribution of the sample mean for samples of size 40

The one exception is for samples of size 1—not a very interesting or informative sample! In this case, the sample mean is equal to the single observation in the sample, and the sampling distribution of the sample mean will look just like the population distribution. For all other sample sizes, the sampling distribution of the sample mean will be less spread out than the population distribution, and the larger the sample size, the smaller the sample-to-sample variability.

So far, we have considered three types of distributions: a population distribution, a sample distribution, and a sampling distribution. Research suggests that students often have difficulty in distinguishing among these types of distributions (see, for example, Chance, delMas, and Garfield [2004] and Lipson [2002]). Understanding the distinction among them is important for correct statistical reasoning. The three graphical displays in figure 1.18 illustrate the difference.

The population distribution, if known, is informative because it tells us about the behavior of some variable in a population. The sample distribution tells us about the behavior of a particular variable in one sample, and if that sample was selected at random, the sample distribution is an approximation of the population distribution. But what does the sampling distribution tell us? Knowing the sampling distribution of a sample statistic allows us to answer the following questions:

Developing Essential Understanding of Statistics for Teaching Mathematics in Grades 6–8 (Kader and Jacobbe 2013) explores tabular and graphical ways to represent distributions.

Population distribution
The distribution of the values of some variable for the entire population. The population distribution describes individual-to-individual variability in the population. In most real-life contexts, the population distribution is unknown.

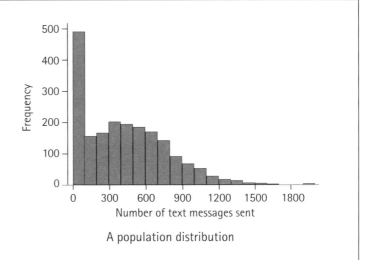

A population distribution

Sample distribution
The distribution of the values of some variable for a sample selected from the population. The sample distribution describes individual-to-individual variability in the sample. We are able to see and work with this distribution.

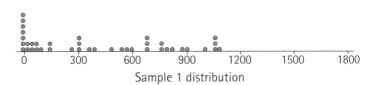

Sample 1 distribution

Sample distribution for one sample of size 40

Sampling distribution
The distribution of the values of a sample statistic (such as the sample mean) for all possible samples of a given size that might be selected from the population. The sampling distribution describes how the value of the statistic varies from sample to sample.

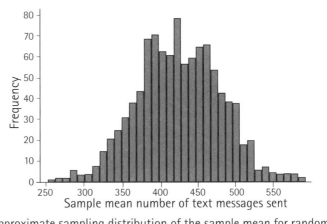

Approximate sampling distribution of the sample mean for random samples of size 40

Fig. 1.18. Population distribution, sample distribution, and sampling distribution

1. How much will the value of a sample statistic tend to differ from one random sample to another?

2. How much will the value of a sample statistic tend to differ from the corresponding population value?

3. On the basis of what we see in a random sample, can we expect to be able to provide an accurate estimate of a population characteristic?

For example, the distribution shown in green in figure 1.17 approximates the sampling distribution of the sample mean for random samples of size 40, drawn from the population that we considered earlier, consisting of 2000 values for the number of text messages sent. In this case, a careful look at this approximate sampling distribution would allow us to answer the questions above as follows:

1. The sample means from random samples of size 40 tend to differ quite a bit from sample to sample. We know this because the approximate sampling distribution has a lot of variability, with observed sample means ranging from about 240 to about 590 text messages.

2. Although sample means tend to center around 420.6, the value of the population mean, some samples resulted in sample means that are quite far from the value of the population mean. Many of the sample means are within about 50 text messages of the actual population mean.

3. Many of the samples resulted in sample means that are within about 50 text messages of the actual population value, and very few samples resulted in a sample mean that is farther than 100 from the value of the population mean. This suggests the degree of accuracy that we can expect if we use the sample mean from a random sample of size 40 as an estimate of the mean of this population.

How would the answers to these questions be different if the sample size were larger? Figure 1.19 shows a histogram of the approximate sampling distribution of the sample mean for random samples of size 100 drawn from the population of numbers of text messages sent. Reflect 1.13 explores the impact of the larger sample size.

Reflect 1.13

Consider the approximate sampling distribution in figure 1.19.

a. How does this distribution differ from the approximate sampling distribution for samples of size 40, shown in figure 1.16?

b. How would you answer the three questions in the text above for the sample mean of random samples of size 100?

The Big Ideas and Essential Understandings

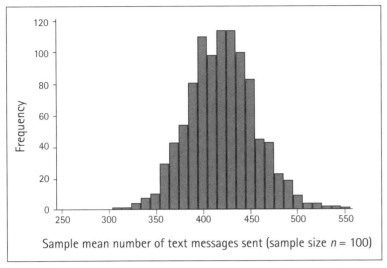

Fig. 1.19. The approximate sampling distribution of the sample mean for samples of size 100, drawn from the population shown in figure 1.11

Although the sampling distribution of a statistic provides important information, you may have noticed that even obtaining an approximation of the sampling distribution by selecting many different random samples is a very tedious and impractical process. Furthermore, in some situations you may not have access to the population to allow you to do repeated sampling. Fortunately, some very general results, based on statistical theory, describe the approximate sampling distributions for many commonly used sample statistics.

For example, in many situations, the approximate sampling distribution of the sample mean, $\bar{x}$, has a predictable form. To investigate this form, let's revisit the text message population shown in figure 1.11. Recall that this population consisted of the values of the numbers of text messages sent by the 2000 students at a high school, and the population mean was 420.6 messages. We have already considered two approximate sampling distributions of the sample mean for this population—one for samples of size 40 and one for samples of size 100. The population distribution and these two sampling distributions are shown again in figure 1.20. Reflect 1.14 explores similarities and differences between the two sampling distributions in the figure.

Both of the approximate sampling distributions appear to be centered at around 420.6, the value of the population mean. The sampling distribution for $n = 40$ is more spread out than the sampling distribution for samples of size 100, and both are less spread out than the population distribution. As previously noted, sample-to-sample variability tends to decrease as the sample size increases.

See Reflect 1.14 on p. 41 before reading further.

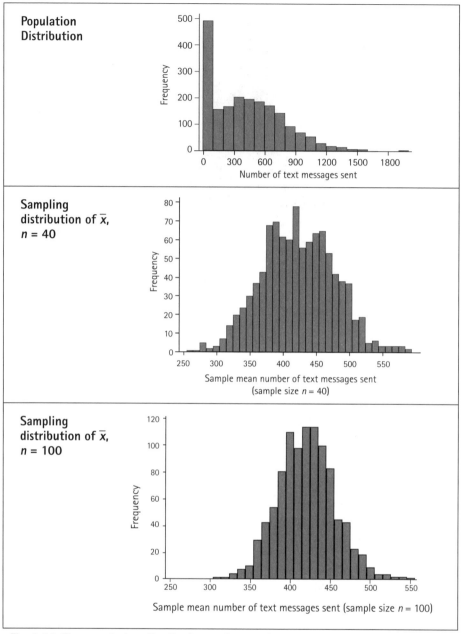

Fig. 1.20. The population distribution and approximate sampling distributions of the sample mean for $n = 40$ and $n = 100$

With respect to shape, both sampling distributions are roughly bell-shaped and symmetric, resembling a normal distribution. You may find this a bit surprising, since the shape of the population distribution is not at all symmetric.

On the basis of these observations, you might wonder about the following:

The Big Ideas and Essential Understandings

> **Reflect 1.14**
>
> Use the histograms in figure 1.20 to answer the following questions:
>
> a. Where do the approximate sampling distributions appear to be centered relative to the population mean of 420.6? Do you think this will always be the case?
>
> b. For which of the two sample sizes is the variability in the sampling distribution greater? How does the variability in the sampling distributions compare with the variability in the population? (Notice that the scale for the population histogram is not the same as the scales for the sampling distributions.)
>
> c. In general, what do you think happens to variability in the sampling distribution of the sample mean as the sample size increases?
>
> d. Are the two sampling distributions similar in shape? How do the shapes of the sampling distributions compare with the shape of the population distribution? Does this surprise you?

- Is the sampling distribution of $\bar{x}$ always centered at the value of the population mean?
- Does the variability of the sampling distribution of $\bar{x}$ always decrease as the sample size increases?
- Is the shape of the sampling distribution of $\bar{x}$ always approximately normal?

Let's look at a few more examples. Figure 1.21 shows three more approximate sampling distributions for random samples selected from the text message population. Reflect 1.15 asks you to think about how you would answer the questions posed above in light of the additional sampling distributions provided in the figure.

> **Reflect 1.15**
>
> Do the approximate sampling distributions in figure 1.21 provide evidence that the answer to any of the questions posed above is no?

Notice that all three of the sampling distributions in figure 1.21 are centered close to 420.6, like the sampling distributions for $n = 40$ and $n = 100$. The variability in the sampling distribution is greatest when the sample size is smallest, so this is also consistent with what we observed for sample sizes of 40 and 100. However, take a careful look at the shapes of the sampling distributions. For the smaller sample sizes of $n = 5$ and $n = 10$, the sampling distribution isn't really symmetric and doesn't have the shape of a normal distribution. But notice that as the sample size increases, the shape

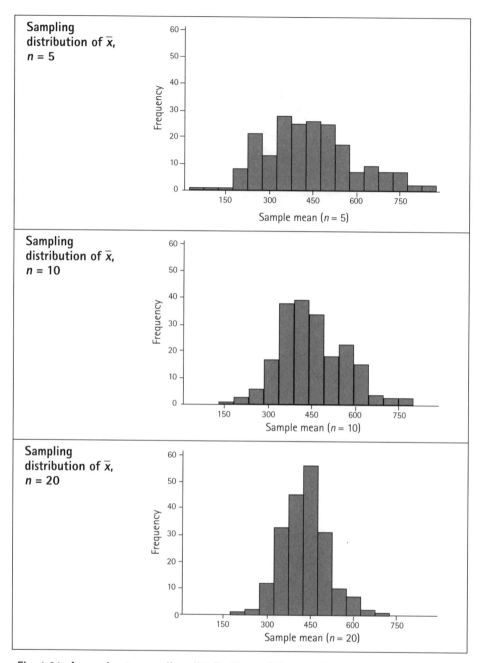

Fig. 1.21. Approximate sampling distributions of the sample mean for $n = 5$, $n = 10$, and $n = 20$

of the sampling distribution of $\bar{x}$ becomes more symmetric and more closely resembles a normal distribution.

These observations are consistent with what statistical theory tells us about the sampling distribution of the sample mean. The foundations of this theory are beyond the scope of this book (for more information, see a statistics text [e.g., Devore 2012]); we simply remark here that if a random sample of size n is selected from a population with mean μ and standard deviation σ, the following three things are known about the sampling distribution of the sample mean $\bar{x}$:

1. The values for $\bar{x}$ from random samples of size n will cluster around the value of the population mean μ.

2. The larger the sample size, the less the values of $\bar{x}$ will tend to differ from sample to sample, and the more tightly they will cluster around the value of the population mean. The standard deviation of the sampling distribution is $\frac{\sigma}{\sqrt{n}}$. Notice that the standard deviation of the $\bar{x}$ sampling distribution depends on the sample size n, and that this standard deviation gets smaller as the sample size gets larger. Because the population standard deviation σ is usually unknown, this standard deviation is estimated by using the standard deviation s calculated from the sample in place of σ in the formula.

3. If the population distribution is normal in shape, the sampling distribution of $\bar{x}$ is also normal in shape. Even if the population distribution is not normal in shape, as long as the sample size n is large enough, the sampling distribution of $\bar{x}$ will be approximately normal in shape. Unless the population distribution is far from symmetric, a sample size of 30 or more is usually considered large enough for the sampling distribution of $\bar{x}$ to be approximately normal.

Statistical theory also provides information about the approximate sampling distributions of some other statistics. Knowing these approximate sampling distributions is the basis for using these statistics to make inferences. How the information provided by a sampling distribution is used in drawing conclusions about population characteristics is a topic that we will explore in connection with Big Ideas 3 and 5.

Big Idea 3

Hypothesis tests answer the question, "Do I think that this could have happened by chance?"

Big Idea 5

Evaluating an estimator involves considering bias, precision, and the sampling method.

The Logic of Hypothesis Testing: Big Idea 3

Big Idea 3. *Hypothesis tests answer the question, "Do I think that this could have happened by chance?"*

Suppose that a company that makes chocolate candy claims that half of the candies in its "Brown Bag Value Pak" are milk chocolate and half are dark chocolate. You reach into the bag and pull out a candy. You don't really like dark chocolate, so you are hoping to get a milk chocolate candy. But the first candy you select is dark. You try again and get another dark chocolate. You shake the bag to mix up the candies and try again. Dark. Are you just unlucky, or are you beginning to doubt the candymaker's claim? Reflect 1.16 asks you to consider when doubt would set in.

Reflect 1.16

How many dark chocolate candies would you have to select in a row before you would "reject" the candymaker's claim? Why did you select this number?

If you continued to get dark chocolates when you selected candies from the bag, at some point you would begin to question the candymaker's claim, and at some later point you would probably become convinced that the claim was false. For many people, the questioning would begin after the third or fourth dark candy, and most people would say they were convinced that the claim was false if the first five or six candies selected were all dark. We might see one, two, or even three dark chocolate candies in a row just by chance, but *if* the candymaker's claim were true, it would be pretty unlikely to see five or six dark candies in a row just by chance. Not *impossible*, but still unlikely—and unlikely enough to convince us that the claim was false.

This example illustrates the reasoning involved in carrying out a hypothesis test. Although it can sometimes be easy to get bogged down in the mechanics of carrying out such a test, the reasoning involved is simple and natural. Five essential understandings support the kind of inferential reasoning that is involved in carrying out a hypothesis test.

Choosing between a null and an alternative hypothesis

Essential Understanding 3a. A hypothesis test involves choosing between two competing hypotheses—the null hypothesis and the alternative hypothesis.

In its simplest form, a hypothesis in statistics is a claim or statement about the value of a population characteristic. Hypothesis testing is a process that allows us to see whether sample data provide support for a particular hypothesis about a population or about the effect of a **treatment** in an experiment. For example, in the chocolate candy example introduced above, we might be interested in using sample data to decide whether we have convincing evidence that the candymaker's claim that half the candies in the Value Pak are milk chocolate is not true.

Although it might seem a bit odd at first, the way in which we decide whether sample data provide convincing evidence in support of a particular hypothesis is to set up a pair of competing hypotheses that include the hypothesis of interest. In the chocolate candy example, the two competing hypotheses might be the following:

$$\pi = .5 \quad \text{and} \quad \pi \neq .5$$

where π is the proportion of milk chocolate candies in the candies used to fill the bags produced by this candymaker. The hypothesis $\pi = .5$ says that half of the candies are milk chocolate—the manufacturer's claim—and the hypothesis $\pi \neq .5$ says that the proportion of milk chocolate candies is different from the candymaker's claim. One way to make a decision between these hypotheses would be to look at *every* candy in the bags. Doing so would be impractical, but if we did this, we would know the actual value of π, and we would know which of these two hypotheses was correct. Unfortunately, we usually must decide between the two competing hypotheses by using data from a sample, which means that we do not have complete information.

A criminal trial is a familiar situation in which a choice must be made between two contradictory claims. The person accused of the crime must be judged either guilty or not guilty. Under the U.S. system of justice, the individual on trial is initially presumed to be not guilty. Only strong evidence to the contrary causes the not-guilty claim to be rejected in favor of a guilty verdict. The burden is thus on the prosecution to *convince* the jurors that a defendant is guilty.

As in a judicial proceeding, in a hypothesis test we initially assume that a particular hypothesis is the correct one. We then consider the "evidence" provided by the sample data and reject this hypothesis in favor of a competing hypothesis *only if* the evidence against the first hypothesis is convincing. The hypothesis that is initially assumed to be true is called the **null hypothesis** and is typically denoted by H_0. The competing hypothesis is called the **alternative hypothesis** and is typically denoted by H_a.

In choosing between the null and alternative hypotheses, we initially assume that the null hypothesis is true, and we reject the null hypothesis in favor of the alternative hypothesis only if the sample data *convinces* us that the null hypothesis is incorrect. This happens when we observe sample data that would have been very unlikely to occur *if the null hypothesis were true*. In this case, we reject the null hypothesis. This is a strong conclusion because we have convincing evidence against the null hypothesis. Figure 1.22 illustrates this process, and Reflect 1.17 explores the distinction between the null and alternative hypotheses.

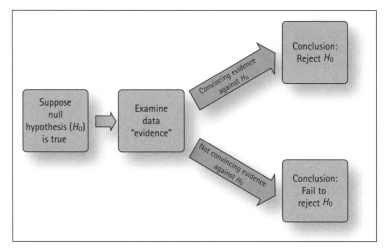

Fig. 1.22. Reaching a conclusion in a hypothesis test

Reflect 1.17

If a criminal trial were viewed as a hypothesis test, which of the following "hypotheses" would be the null hypothesis, and which would be the alternative hypothesis?

The defendant is guilty.

The defendant is not guilty.

In the trial analogy, "The defendant is not guilty" would be the null hypothesis because it is the hypothesis that is initially assumed to be true. We would reject this hypothesis only if the evidence against it were convincing. If we had any doubt, we would not reject the "not guilty" hypothesis. These two possible decisions—rejecting the null hypothesis and not rejecting the null hypothesis—are not equivalent, a situation that Reflect 1.18 investigates.

The Big Ideas and Essential Understandings

> **Reflect 1.18**
>
> Rejecting the null hypothesis is a strong conclusion. Do you think that failing to reject the null hypothesis is also a strong conclusion? Why do you think this?

If, on the basis of the sample data, we are not convinced that the null hypothesis is false, then we fail to reject the null hypothesis. It is important to note that this does *not* mean we have strong or convincing evidence that the null hypothesis is true. We can say only that we are not convinced that the null hypothesis is false. This is different from saying that we are convinced that it is true.

For example, in the chocolate candy example, if we reject the null hypothesis that $\pi = .5$, we can say that there is convincing evidence that the proportion of milk chocolate candies in the bags produced by this candymaker is not .5 and that the candymaker's claim is wrong. However, if we fail to reject the null hypothesis $\pi = .5$, which might happen, say, if 3 out of 4 of the first chocolates chosen were dark, we do so not because we were convinced that the candymaker's claim is true. We would be saying only that we were *not convinced* that the proportion of milk chocolate candies was different from .5. The reason that we were not convinced is that the difference between the proportion of milk chocolate candies in the sample and the claimed value of .5 can be explained by the chance differences that will occur from sample to sample when the candymaker's claim is true.

Determining the alternative hypothesis

Essential Understanding 3b. The alternative hypothesis is determined by the statistical question of interest.

A statistical hypothesis test can demonstrate strong support only for the alternative hypothesis. Because of this, if the goal of a study is to decide whether sample data support a particular claim about a population, we will want this claim to determine the alternative hypothesis. The alternative hypothesis will usually have one of the following three forms:

H_a: population characteristic > hypothesized value

H_a: population characteristic < hypothesized value

H_a: population characteristic ≠ hypothesized value

The study context and the statistical question of interest determine the hypothesized value and whether "greater than," "less than," or "not equal to" (>, <, or ≠) is the appropriate relation between this hypothesized value and the population parameter. This hypothesis is formulated *before* looking at the sample data because its selection is determined by the statistical question that we are trying to answer and should not be influenced by the observed data.

The null hypothesis is usually stated as a claim of equality (=). However, sometimes the null hypothesis is written as a "less than or equal to" (≤) relationship when the inequality in the alternative hypothesis is stated as a "greater than" (>) relationship. Or you might see the null hypothesis written as a "greater than or equal to" (≥) relationship when the inequality in the alternative hypothesis is stated as a "less than" (<) relationship. However, the "equal to" (=) case is always included as part of the null hypothesis.

To see why it is reasonable to use a statement of equality in the null hypothesis, suppose that the alternative hypothesis for a particular test is $H_a: \mu > 750$. The null hypothesis would usually be written as $H_0: \mu = 750$, rather than $H_0: \mu \leq 750$. This allows us to focus on the boundary value of 750. Rejecting the null hypothesis of $\mu = 750$ in favor of the alternative hypothesis of $\mu > 750$ then means that we were convinced that the actual mean is greater than 750. It then must also be the case that we were convinced that the mean is greater than 749 or than 748 or than any value smaller than 750.

An example may help to illustrate how the selection of H_0 (the claim initially believed true) and H_a depend on the objectives of a study. Suppose that a medical research team has been given the task of evaluating a new treatment for a particular disease. Consider the following two scenarios:

> Scenario 1: The current standard treatment is considered reasonable and safe by the medical community, has no major side effects, is not expensive, and has a known recovery rate of .9 (90% of patients recover).

> Scenario 2: The current standard treatment sometimes has serious side effects, is costly, and has a known recovery rate of .2 (20% of patients recover).

In an investigation of the new treatment under the first scenario, the statistical question of interest would probably be, "Does the new treatment have a higher recovery rate than the standard treatment?" Unless the investigation found convincing evidence that the new treatment had a higher recovery rate, it is unlikely that current medical practice would change. With π representing the true recovery proportion for patients who received the new treatment, the following hypotheses could be tested:

$$H_0: \pi = .9 \quad \text{versus} \quad H_a: \pi > .9$$

In this case, rejecting the null hypothesis would indicate convincing evidence of a higher recovery rate for the new treatment. Reflect 1.19 asks you to consider how you might modify the hypotheses in the very different case of scenario 2.

The Big Ideas and Essential Understandings

> **Reflect 1.19**
>
> Suppose that you were investigating a new treatment under scenario 2. Would you choose the same null hypothesis and alternative hypothesis as in the case of scenario 1? If not, what hypotheses would you choose?

In scenario 2, the current standard treatment does not have much to recommend it. The new treatment might be considered preferable because of cost or because it has fewer or less serious side effects, as long as the recovery rate associated with it is no worse than that associated with the standard treatment. Here, researchers might decide to test the following:

$$H_0: \pi = .2 \quad \text{versus} \quad H_a: \pi < .2$$

On the one hand, if they reject the null hypothesis in favor of the alternative hypothesis, the new treatment will not be put forward as an alternative to the standard treatment, because there is strong evidence that it has a lower recovery rate than the standard treatment. On the other hand, if the researchers do not reject the null hypothesis, they are able to conclude only that there is not convincing evidence that the recovery rate for the new treatment is lower than that for the standard. This is *not* the same as saying that they found evidence that the new treatment is as good as the standard treatment. If medical practice embraces the new treatment, it will not be because it has a higher recovery rate, but rather because it costs less or has fewer side effects and there is no convincing evidence that it has a lower recovery rate than the standard treatment.

Convincing evidence against the null hypothesis

Essential Understanding 3c. The null hypothesis is rejected in favor of the alternative hypothesis if the sample data provide convincing evidence against the null hypothesis.

Let's return to the chocolate candy example and the two hypotheses between which we are interested in deciding:

$$H_0: \pi = .5 \quad \text{and} \quad H_a: \pi \neq .5,$$

where π is the proportion of candies that are milk chocolate. What would it take to convince us that the null hypothesis, H_0, is false? Earlier, we said that if we sampled 5 or 6 candies and all of them were dark chocolate, we thought that most people would consider that result convincing. But what if we sampled 10 candies and found that only 3 of them were milk chocolate? Should we consider that convincing evidence against the null hypothesis that $\pi = .5$? To make this decision, we need to evaluate whether it would have been surprising to observe only 3 milk chocolate candies in the sample of 10 candies *if the population of all candies really did consist of half milk chocolate and half dark chocolate.*

One way to investigate this proposition is to try sampling from a population of candies that we *know* contains half milk chocolate candies. Reflect 1.20 asks you to think about how we might simulate drawing samples from such a population.

Reflect 1.20

Consider a population that consists of the digits in a **random number table** or a collection of random digits generated by a random number generator on a graphing calculator. How might this population of random digits be viewed as a hypothetical population of candies, half of which are milk chocolate?

We could use a collection of random digits to simulate a population of candies, with each digit in the collection simulating one candy. If we thought of even digits as milk chocolate candies and odd digits as dark chocolate candies, we would have a simulated population of candies, half of which were milk chocolate.

Now consider taking a random sample of 10 from this population. This sample would consist of 10 random digits. Reflect 1.21 explores the use of a simulation to learn about what might be expected in a sample if the null hypothesis were true.

Reflect 1.21

Suppose that the following 10 random digits represent a sample of 10 candies from a population in which half of the candies are milk chocolate:

4 0 6 9 7 1 4 5 9 3

If even digits represent milk chocolate candies, how many milk chocolate candies are in this simulated sample of 10 candies?

➡ **Essential Understanding 2c**
The sampling distribution of a sample statistic describes how the value of the statistic varies from sample to sample.

The 10-candy sample corresponding to the random digits in Reflect 1.21 has 4 milk chocolates. If we repeated this process many times, each time using a different set of random digits, we would get a sense of what to expect when taking samples of 10 candies from a population for which we *know* half of the candies are milk chocolate.

Figure 1.23 is a dot plot of the number of milk chocolate candies observed in a sample of 10 candies for 100 different random samples from a population with half milk chocolate candies. Thus, this figure is an **approximate sampling distribution** for the statistic that counts the number of milk chocolate candies in a sample of 10 chocolates drawn from a population in which exactly half the candies are milk chocolate (see the discussion of Essential Understanding 2c). Reflect 1.22 explores the use of this distribution to decide whether any particular sample outcome is unusual.

The Big Ideas and Essential Understandings

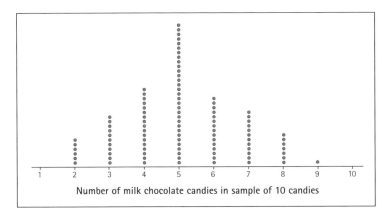

Fig. 1.23. The number of milk chocolate candies in samples of size 10, drawn from a population of candies, half of which are milk chocolate

Reflect 1.22

On the basis of the dot plot in figure 1.23, can you say that seeing only 3 milk chocolate candies in a random sample of size 10 is surprising for a population that really does consist of half milk chocolate candies? Would you consider 3 milk chocolate candies in a random sample of size 10 *convincing* evidence against the null hypothesis H_0: π = .5?

Looking at figure 1.23, we can see that finding as few as 3 milk chocolate candies when π = .5 would not be surprising. In fact, 17 out of the 100 samples included 3 or fewer milk chocolate candies. If π = .5, a sample with 3 or fewer milk chocolate candies would occur about 17% of the time, simply as a result of chance differences that we expect to occur from sample to sample. Because sampling variability is a plausible explanation for why we might see only 3 milk chocolate candies in a sample of size 10, we would not be convinced that H_0: π = .5 was false, and we would not reject this null hypothesis. Reflect 1.23 invites you to take the next step.

Reflect 1.23

On the basis of figure 1.23, what number of milk chocolate candies would you consider surprising in a sample of 10 from a population of candies that consists of half milk chocolate candies?

Notice that none of the samples used to construct the dot plot in figure 1.23 had 0 or 1 milk chocolate candy. On the high end, no sample had 10 milk chocolate candies, and only one sample had 9. Because these events occur infrequently, we might consider it surprising to see 1 or fewer milk chocolate candies, and also surprising to find 9 or more milk chocolate candies, in a sample of size 10 when $\pi = .5$. If we observed one of these results, we could reject the null hypothesis. Although we would not have proof that the null hypothesis is false, we might be convinced that it is false because we saw something that, while not impossible, is very unlikely to occur if the null hypothesis is actually true. The process just illustrated for evaluating whether sample data are surprising if the null hypothesis is actually true is formalized in two remaining essential understandings associated with Big Idea 3.

A measure of surprise

Essential Understanding 3d. *The p-value measures surprise.*

The *p*-value in a hypothesis test is defined as the probability of observing sample data as extreme as, or more extreme than, the data observed in the sample, *given that* the null hypothesis is true. Notice the conditional nature of this statement: the *p*-value is calculated under the assumption that the null hypothesis is true. Defined in this way, the *p*-value can be thought of as a measure of inconsistency between the claim made in the null hypothesis and the observed sample data. Reflect 1.24 explores how the *p*-value is used in reaching a decision in a hypothesis test.

Reflect 1.24

Is a small *p*-value an indication that the null hypothesis should be rejected or that it should not be rejected?

If the *p*-value is small, this means that if the null hypothesis is true, it is very unlikely—in other words, the probability is very small—that we would see something like what we observed in the sample. This indicates that the sample is inconsistent with what we would expect to see if the null hypothesis were true, and we interpret this as evidence against the null hypothesis.

By contrast, if the *p*-value is not small, this means that if the null hypothesis is true, it would not be unusual to observe something like what we see in the sample. Although this does not *prove* that the null hypothesis is true, it means that the observed sample is consistent with what we might expect to see simply as a result of the chance variability that will occur from sample to sample when the null hypothesis is true. There is then no reason to reject the null hypothesis.

So how are *p*-values actually determined? One approach is to estimate a *p*-value by carrying out a simulation like the one described in the discussion of Essential Understanding 3c for the chocolate candy example. However, although carrying out such a simulation is helpful in understanding what a *p*-value measures, it can

Essential Understanding 3c
The null hypothesis is rejected in favor of the alternative hypothesis if the sample data provide convincing evidence against the null hypothesis.

be a very tedious process. This is where knowing something about sampling distributions—the focus of Essential Understanding 2c—can come to the rescue by providing another way to estimate *p*-values. To see how an understanding of the sampling distribution of a sample mean can provide insight into estimating a *p*-value, consider the following example.

An article in the journal *Obesity* (Dumanovsky et al. 2009) describes a study in which 3857 adult customers, approached as they entered a restaurant in one of three hamburger chains, agreed to provide their receipts when exiting. Researchers used the receipts to determine what items the customers had ordered, and then they computed the total number of calories in the meal. A website on healthy dining (www.healthydiningfinder.com) recommends a target of 750 calories for lunch, based on a diet of 2000 calories per day. Researchers were interested in determining whether the mean number of calories in a lunch at a restaurant in any of the three chains was greater than the recommended value of 750 calories.

Of course, we would probably want to know more about the study, such as how the 3857 people were selected and whether many people who were approached refused to participate. But for this discussion, let's assume that it is reasonable to regard this sample as representative of the population of people who eat lunch at restaurants in the three hamburger chains. The researchers decided to test the following hypotheses:

$$H_0: \pi = 750 \quad \text{and} \quad H_a: \pi > 750$$

The article reported that for the sample of 3857 people, the sample mean number of calories consumed was 857, and the sample standard deviation was 677. The sample mean is certainly greater than 750, but the key question here is whether this statistic is consistent with what we would expect to see simply as a result of chance differences from sample to sample *when the population mean is really 750*. To decide whether a sample mean of 857 would be surprising if μ really is 750, we need to know what values for sample means would be expected if the population mean is 750. Fortunately, this is exactly the kind of information we can get from considering the sampling distribution of the sample mean.

Recall from our discussion of Big Idea 2 that we know the following three things about the sampling distribution of the sample mean $\bar{x}$:

1. *If the population mean really is 750*, the $\bar{x}$ values from samples selected at random from the population will be centered around 750.

2. The standard deviation of the sampling distribution of $\bar{x}$ is approximately $\dfrac{s}{\sqrt{n}} = \dfrac{677}{\sqrt{3857}} \approx 10.9$.

> **Essential Understanding 2c**
> The sampling distribution of a sample statistic describes how the value of the statistic varies from sample to sample.

Big Idea 2

Distributions describe variability.

3. Because the sample size is large, the sampling distribution of $\bar{x}$ is approximately normal in shape.

So, if the population mean really is 750, the $\bar{x}$ values from random samples of size 3857 should (approximately) follow a normal distribution with mean 750 and standard deviation 10.9.

Because we know that the sampling distribution is approximately normal, we can make use of a result called the *empirical rule*. The empirical rule states that for normal distributions—

- about 68% of the values will be within 1 standard deviation of the mean;
- about 95% of the values will be within 2 standard deviations of the mean; and
- about 99.7% of the values will be within 3 standard deviations of the mean.

The sample mean in this study was 857, or 107 greater than the hypothesized value of 750. Because the standard deviation of the sampling distribution of the sample mean is approximately 10.9, the observed sample mean is approximately $\frac{107}{10.9} = 9.82$ standard deviations above 750. This is not something that we would expect to see in a normal distribution. It is very unlikely that we would see a sample mean that is more than 3 standard deviations above 750, and the chance of seeing one that is almost 10 standard deviations above 750 is essentially 0. This result corresponds to the statement that the *p*-value for this test is approximately 0. And this tells us that if the null hypothesis is true, it is *extremely* unlikely that we will see a sample mean as large as 857. Thus, we have strong evidence against the null hypothesis, which we reject in favor of the alternative hypothesis.

There are formal procedures for carrying out hypothesis tests of various types, but the reasoning behind all tests is essentially the same as in the example considered here. Understanding this logic and the interpretation of *p*-values can be very valuable in other hypothesis-testing situations.

A risk of error

Essential Understanding 3e. *Hypothesis tests do not always lead to a correct decision.*

In hypothesis-testing situations, we use partial rather than complete information to determine whether we have strong support for a particular hypothesis about a population characteristic. Consequently, we run the risk of an incorrect conclusion. On the basis of the available data, we might conclude that there is convincing evidence against a hypothesis that is in fact true. Or we might decide on the basis of the sample data that we are not convinced that a hypothesis should be rejected when in fact it is not true and should have been rejected. Consider the following example.

Physicians in England wished to test whether listening to music helped calm patients about to undergo an uncomfortable medical procedure (El-Hassan et al. 2009). Using a validated measure of anxiety, researchers recorded the change in anxiety levels for each person in a sample of patients. Although the patients were not a random sample, they were randomly assigned to either a treatment group (music played before the procedure) or a control group (no music). The researchers concluded that the control group experienced "no difference" in anxiety, whereas the treatment group experienced a "significant reduction in anxiety."

Let's assume that the only possible explanation for an observed difference between the control and the treatment groups' change in anxiety levels is that either (a) the treatment really lowers anxiety or (b) the observed difference is due to chance. If the study is well designed and the numerical measures of anxiety accurately reflect the patients' anxiety levels, then the assumption that (a) is the explanation may be sound. Still, this conclusion could be incorrect—that is, the decision to reject the null hypothesis in favor of the alternative hypothesis could be a mistake on the part of the researchers. Repeating the study on even the same group of patients is certain to produce slightly different numerical results, in part because anxiety levels, reactions to music, and reactions to medical procedures all have the potential to be influenced by a variety of factors beyond the researchers' control. It is possible, therefore, that music has no effect, and the observed difference was due merely to chance, even though for these particular samples, the difference between control and treatment groups seem large enough for researchers to conclude that anxiety was affected by listening to music. A different type of error would occur if the researchers concluded that there was no evidence of a difference between the two groups when there was a real difference, which, because of variability in the data, the researchers were unable to detect.

Acknowledging the risk of error is an important part of carrying out a hypothesis test. The **significance level** of a hypothesis test is defined as the probability of rejecting the null hypothesis, given that the null hypothesis is true. The significance level is thus a **conditional probability**; in the study of the effect of listening to music on anxiety, it is the probability that the researchers will conclude (incorrectly) that listening to music reduces anxiety *if* in fact it does not.

The significance level can be used as one measure of the risk of an incorrect decision in a hypothesis test. Although the significance level does not tell us whether a particular decision to reject or fail to reject the null hypothesis is correct, it does provide information about the risk of rejecting a true null hypothesis for the particular method used to reach a conclusion. We do not know the probability that the medical researchers' conclusions are incorrect for the particular study described. However, because they reported a significance level of .05, we do know that they used a method that declares an effect by mistake in 5% of all random samples taken in situations where there is no real effect. We also know that, all other things being equal, we would generally prefer this test over a test that makes this mistake 10% of the time.

The significance level of a test is determined before the test is carried out. The chosen significance level can be achieved if we reject the null hypothesis whenever the associated *p*-value is less than the specified significance level. For example, suppose that the significance level for a test is chosen to be .05. If we reject the null hypothesis only when we see something that would have occurred by chance less than 5% of the time, given that the null hypothesis is true (equivalently, *p*-value < .05), our risk of incorrectly rejecting the null hypothesis is controlled at 5%.

Of course, incorrectly rejecting the null hypothesis is only one type of possible error. The other type of error is not rejecting the null hypothesis when it should have been rejected. Unfortunately, it is not as easy to quantify this risk, because it depends not only on the sample size and the significance level of the test, but also on the actual value of the population characteristic.

To consider an example, let's revisit the calorie study, for which the null and alternative hypothesis in a test were

$$H_0: \mu = 750 \quad \text{and} \quad H_a: \mu > 750.$$

Suppose that the mean number of calories consumed at lunch is actually greater than 750. In other words, the null hypothesis should be rejected. The risk of not rejecting the null hypothesis (an incorrect decision) depends on the actual value, μ, of the mean number of calories consumed by the population. It is more likely that we will make a correct decision to reject the null hypothesis if μ is really 800 than if it is really 751. Because we do not know the actual value of μ (if we did, there would be no reason to carry out the test!), we can't easily quantify the risk of incorrectly failing to reject the null hypothesis.

This is one reason why the "fail to reject" decision in a hypothesis test is not a strong conclusion. We are not convinced that the null hypothesis is incorrect, so we do not reject it. But we also do not know the chance that this is an incorrect decision. For more background on assessing the risks of an incorrect decision in hypothesis tests, consult a statistics text (e.g., Devore 2012).

The Importance of the Data Collection Method: Big Idea 4

Big Idea 4. *The way in which data are collected matters.*

The statistician David Moore famously defined data as "numbers in context." Partly, this means that data are not "pure" numbers, as in mathematics, but include units of measurement and, by implication, measurement error. But this "context" also includes the method used to collect the data and the need that motivated the collection of the data in the first place.

The method of collection matters because it limits—or enhances, if you are an optimist—the types of inference that you can make on the basis of data. For example, the Pew Research Foundation did a survey that reported that about one-third of the population sends more than 3000 text messages each month. If you are over 18 years of age, you will probably be incredulous about this figure until you learn that the researchers sampled only U.S. teenagers, so the "population" to which they refer does not include adults.

Data collection is also important because statisticians earn their money not just through estimation and prediction, which, when you think about it, anyone can do, but through quantifying potential errors in their estimations and predictions. To predict that a candidate will get 55% of the vote is a statement that anyone can make. But for someone to quantify the confidence with which he or she makes a statement that the candidate will get between 52% and 58% of the vote requires statistics.

These quantifications of errors are based on knowledge of the sampling distributions of our statistics, and these, in turn, require that the data be collected through some type of random process. Without random sampling or random assignment, it is difficult, if not impossible, to quantify uncertainty. Four essential understandings about the importance and implications of data collection support Big Idea 4.

Observational studies and controlled experiments

Essential Understanding 4a. *Observational studies, including surveys, provide information about the characteristics of a population or sample, whereas controlled experiments provide information about treatment effects.*

The kind of things that we learn from observational studies are different from those that we learn from controlled experiments, so distinguishing between the two is important. Although there are many different types of observational studies, and learning to design a controlled experiment could easily occupy at least one entire year-long course, some identifying features can be used to distinguish between these two types of studies.

Controlled experiments are investigations in which researchers have some level of control over the **treatments** given to subjects in the study. Most experiments are designed to measure the effect of a treatment—or to determine whether the treatment has any effect at all. A treatment might consist of administering a drug or exploring the effects of different doses of a drug. Alternatively, the treatment might be an experience: listening to Mozart while solving a puzzle. The goal of an experiment is to measure the effect of this treatment on a response: is the disease cured or mitigated? Is the puzzle solved more quickly? A good study should include a comparison group, sometimes also called a *control group*. We cannot conclude that people who solved the puzzle while listening to Mozart did so more quickly than those who did not listen to Mozart unless we have observed a "no Mozart" comparison group. (Some studies will compare several different groups, comparing, for example, different dosage levels of a drug.)

For the sake of simplicity, we can think of experiments as attempts to explore the association between two variables. The first of these variables is the *treatment variable*, which records, for each subject, the treatment received. This might be a categorical variable that simply records whether the subject received the treatment (listened to Mozart) or the control (listened to no music). Or the treatment variable might, for example, be a variable that records the dosage of some medication that the subject received. The second variable is the *response variable*, which is the variable that researchers are interested in affecting. For example, this variable might be a measure of how quickly the subject solved the puzzle or the subject's health at some time in the future.

In a controlled experiment, researchers determine the value of the treatment variable for each subject. For example, in the simplest of studies, in which there are only two values for the treatment variable (because subjects either receive the treatment or are placed in a comparison group), researchers will assign subjects to one group or the other. To avoid introducing **bias** in estimates of the effect of the treatment, researchers should use a randomization mechanism to assign subjects to treatments. If, for instance, physicians made this assignment themselves, they might assign the sickest subjects to the treatment that they believed would be most effective. Instead, to avoid bias, researchers might flip a coin to determine which subjects will be assigned to the treatment group and which ones will be assigned to the comparison group.

Observational studies stand in contrast to controlled experiments in that the researchers do not assign subjects to treatments.

Sometimes it is impossible (or prohibitively expensive) to conduct a controlled experiment, or sometimes doing so may be unethical. For example, one cannot assign subjects to a treatment known to be harmful, such as smoking cigarettes. Thus, researchers merely "observe" the "assignments" that subjects have determined for themselves or that have been determined by forces beyond anyone's control. For example, we might study whether female or male restaurant servers get better tips. Although we can control a number of factors (the type of restaurant, the prices on the menu, the time of day), we cannot assign a gender to the server. Most likely, we would perform an observational study in which we recorded data on a large number of male and female servers at similar restaurants at similar times of day. (By using "similar" restaurants, we minimize variation in our response variable that might cause inappropriate comparisons if, say, tipping behavior—and server behavior—are very different at mid-range and high-end restaurants.) Reflect 1.25 explores why controlled experiments are sometimes difficult to perform and observational studies may be conducted instead.

> **Reflect 1.25**
>
> In recent years, we have had frequent news stories about reports suggesting that talking on a cell phone "excessively" may increase the risk of brain cancer. Would you expect the studies behind these claims to be controlled experiments or observational studies? Why?

Controlled experiments are quite expensive, in part because they require a large degree of control over the subjects. To determine whether ten years of heavy cell phone use leads to a higher cancer incidence than ten years of no cell phone use requires enforcing some pretty strict terms of behavior for subjects. It probably could be done, but the costs would be enormous, not to mention the possible ethical issues associated with a treatment that some believe might cause brain cancer. For a number of reasons, then, studies on the risk of cell phone use are more likely to be observational studies, based on observing people who are already engaging in heavy, light, or no cell phone use.

If controlled experiments are so expensive, why do them at all? The reason is that they can reveal cause-and-effect relationships between variables. Our discussion of Essential Understanding 4b will explore the how and why, but in a nutshell, it amounts to this: if we determine that an association exists between the treatment and the response variables in a controlled experiment that is well designed, then we can also conclude that changes in the treatment variable *cause* changes in the response variable. The devil is in the details, naturally, so the simple two-word phrase "well designed" is quite important.

Essential Understanding 4b Random assignment in an experiment permits drawing causal conclusions about treatment effects and quantifying the uncertainty associated with these conclusions.

Observational studies, by contrast, can reveal various characteristics of a population, but they do not allow us to draw cause-and-effect conclusions. We may find an association between the gender of a server and the amount of money left as a tip, for example, but we cannot conclude that this association exists *because of* the server's gender. Similarly, we might discover that the group of people who use cell phones frequently do get a certain type of cancer at a greater rate over a ten-year period than those who use their cell phones less frequently, but we cannot conclude that the cell phones are the culprits, because we cannot eliminate many other potential explanations for this association.

The most common type of study that we see in public life is the *survey*. Reflect 1.26 explores how to classify surveys with respect to our data collection paradigm.

Reflect 1.26

In a survey, researchers randomly choose participants and ask them questions. The results of the survey are used to estimate characteristics of the population—for instance, the proportion who will vote for a particular candidate for president, or the proportion who will refinance their home in the next month. Are surveys controlled experiments or observational studies?

Surveys occupy something of a gray area in the controlled experiment–observational study dichotomy. On the one hand, the researcher is far from passive, often expending great effort in choosing and contacting participants, although for a well-designed survey this choice needs to be based on random selection. The researcher also poses questions directly to the survey participants. However, surveys usually ask about things that the participants have already done or intend to do or about characteristics that they already possess. Research surveys are designed to uncover associations between group membership and behavior—for example, whether the proportion of regular churchgoers voting for a particular candidate will be greater than the corresponding proportion of non-churchgoers. But these studies do not assign the respondents to the groups under examination. For this reason, surveys are often classified as observational studies. We should note, however, that marketers (and politicians) sometimes do use surveys to affect the respondents' behavior, perhaps by making them aware of a product (or candidate or issue) that they knew nothing about before the pollster called them.

Random assignment

Essential Understanding 4b. Random assignment in an experiment permits drawing causal conclusions about treatment effects and quantifying the uncertainty associated with these conclusions.

Causal inference, the technical term for concluding whether or not changing one variable results in a change in the other, is somewhat foreign territory for most mathematicians. Most statistical inference is based on using measurements from a sample to draw conclusions about a population. But controlled experiments are not about extending results to a larger group of *subjects* (at least, not right away), but instead are about understanding whether an observed difference in the response variable was caused by the differences in the treatments.

The guiding principle behind designing a controlled experiment is to make the treatments and control groups as much alike as possible in every way except, of course, for the differences imposed by the treatment. Well-designed controlled experiments often include the following features:

- At least one comparison group, often called a *control group*

- Random assignment of subjects to treatment groups

- A placebo—a treatment with no effect—administered when appropriate, so that subjects in all groups, even the control group, are treated similarly

- Double-blinding, when appropriate, so that neither the subjects nor the people measuring the response know which subjects are in which group.

To illustrate why these features are important, let's consider a simple case with a single treatment whose effectiveness we wish to determine. For example, is applying duct tape an effective way to eliminate a wart (CBS News 2009)? Imagine a simple treatment in which a wart is covered by a piece of duct tape for a fixed period of time. The response variable is the change in the size of the wart, measured after that period of time.

Our first step is to recruit subjects, and we immediately become aware that because of the lack of a national registry of wart sufferers, we have no way of getting a random sample of warts. We instead contact local doctors and dermatologists and ask them to refer their wart-suffering patients to us. We find that we will need a large number of subjects because we quickly realize that there is great variability in both people and their warts. Once we have recruited a sufficient number of subjects and received their informed consent, we realize that we cannot simply give the treatment to everyone, because if we do, we will never know what would have happened if we did not give the treatment. Hence, we need a control group composed of wart sufferers who will not receive any treatment for their warts. Reflect 1.27 highlights the importance of random assignment.

> ### Reflect 1.27
> What sort of biases might we expect in this study of duct tape and warts if researchers personally assigned individual subjects to receive either the duct tape treatment or the control treatment?

Our next step is to assign our subjects to either the treatment or the control group. If we make this decision ourselves, we might, out of sympathy, assign the subjects with the biggest warts or the worst warts to the treatment group, reasoning that they need help the most. But this could bias our estimates of the effect of using duct tape, since these warts might be the least responsive to treatment. We could let the subjects determine their own group assignments, but then we could not be sure that some other difference between the groups did not account for any changes in the response variable.

Confounding factors are variables that provide possible alternative cause-and-effect explanations of associations. A confounding factor must account for why subjects in the treatment group are different from those in the control group and must also explain why they tend to differ on the response variable as well. For example, it is possible that younger subjects are more adventurous and as a result, if given a choice, would be more likely to choose the duct tape group over the control group. It is also possible that the way in which warts respond to treatment depends on the subject's age. If so, then age is a confounding factor.

Identifying potential confounding factors is not easy and requires some knowledge of the context in which the data were collected. Reflect 1.28 provides an opportunity to practice identifying potential confounding factors.

> ### Reflect 1.28
> On the basis of observational studies, early twentieth-century researchers discovered a strong, positive association between the number of cigarettes smoked and the risk of lung cancer. R. A. Fisher, a prominent statistician, said that this did not prove that smoking caused lung cancer, and he suggested a possible confounding factor. Can you suggest a potential confounding factor and explain why it might introduce confusion in the interpretation of the studies' results?

Fisher, who was a pioneer in genetic research, proposed that genetics might be a confounding factor. Some people, he suggested, might have a gene that results in a tendency toward addiction to the chemicals in cigarettes. This same gene, regrettably, might also cause lung cancer. Thus, genetics was a potential confounding factor. Later studies compared identical twins, in which one twin in each pair smoked and the other did not. Because the smokers got lung cancer at a greater rate than their nonsmoking siblings, researchers were able to eliminate genetics as a confounding factor.

One drawback of observational studies is that we can never guarantee that all confounding factors have been eliminated. An observational study can uncover and quantify associations, but the possibility always exists of a confounding factor that explains an association. Over time, researchers can identify potential confounding factors and possibly design studies that eliminate them from consideration.

Random assignment of subjects to treatment groups allows us to rule out confounding factors because if the sample sizes are large enough, the distribution of any potential confounding factor will be about the same in all treatment groups and in the control group. All groups will have about the same proportion of women, for example, about the same distribution of ages, and about the same distribution on variables that we might not even have thought of or cannot easily observe. In other words, we can be reasonably certain that the only meaningful difference between the groups is that they received different levels of the treatment or no treatment at all.

Once we begin treating the subjects, we still need to recognize the possibility that we will induce differences in the groups. People often respond just to the idea of getting treatment (called the *placebo effect*), so it is important that subjects not know, whenever possible, whether they are in the treatment or the control group; technically, this is called a *blind*. If the subjects are blinded, they won't behave differently.

Ensuring that all subjects behave the same, regardless of their assignment to the treatment or the control group, is crucial. For instance, in a trial to determine whether a vaccine is effective at preventing the flu, we could imagine that subjects who believed that they were inoculated might take fewer precautions against the flu than subjects who believed that they had no other defense. A *placebo*—a sham treatment—is often used so that all subjects do not know whether they are receiving the treatment or the placebo. Of course, for a placebo to be useful, the subjects must not know that they have received one. Sometimes, particularly in medical contexts, the use of a placebo is not ethical, and so instead, subjects in the control group receive a standard treatment (rather than the experimental treatment).

One other complication is that the people measuring the response might treat subjects differently if they know who is receiving a placebo and who is receiving a treatment. Many medical diagnoses, for example, require some subjective judgment, and physicians must take into account many factors that are difficult to define. If a physician knows that the patient is receiving a potentially promising treatment, he or she might be inclined to alter the diagnosis. A *double-blind* study attempts to prevent this by ensuring that neither the subjects nor the people measuring their responses know who is getting the treatment and who is getting the placebo. This is easier said than done, of course, but in many studies, a third party keeps track of patient identities and treatment assignments, revealing the subjects' assignments only at the end of the study.

Extending conclusions from a controlled experiment to a larger population can be problematic if the subjects did not constitute a random sample from that population. We might conclude that a treatment was effective for this particular sample of subjects, but we cannot guarantee that results will apply to another group. Medical studies are rarely based on representative samples. In fact, almost the opposite is true. Many medical studies apply "exclusion criteria" to bar participants who have complicating medical conditions, who might respond poorly to treatment, or for whom treatment could be life threatening. For this reason, medical therapies are sometimes less effective in the general population than they were on subjects in testing.

Random selection

Essential Understanding 4c. *Random selection tends to produce samples that are representative of the population, permitting generalization from the sample to the larger population and also allowing the uncertainty in estimates to be quantified.*

Essential Understanding 2a
A population distribution describes variability in the values that make up a population.

Big Idea 5
Evaluating an estimator involves considering bias, precision, and the sampling method.

Common sense tells us that if a sample is not representative of the population, then we must be suspicious of attempts to generalize findings from that sample to the population. A study about teenagers' texting habits cannot be reliably extended to the adult population, for example.

One reliable way to collect a sample that is representative of a population is to select members of the sample at random from that population. The sample will not be identical to the population, of course. In fact, for small samples, differences between the population and sample might be large (see our earlier discussion of Essential Understanding 2a). Random samples are useful because the ways in which they differ from the population are predictable and quantifiable, in some sense. We can also show that for some **estimators**, the variability is less for larger samples, and so the estimated values tend to come closer to the corresponding population value as the sample size increases (estimators are the focus of Big Idea 5; see our discussion for more details).

The Big Ideas and Essential Understandings

The formulas for standard errors and means of estimators that appear in most textbooks assume a "random sample." But there are many different ways to collect a random sample. In introductory statistics, we almost always assume two things: (1) members of the sample were selected at random from the population without replacement, and (2) the population is quite large in relation to the sample size (our discussion of Essential Understanding 5h offers more details).

Selection without replacement means that once a member has been selected for the sample, that member cannot be selected again, and it also means that all samples of the same size are equally likely to be chosen. In real life, collecting such a random sample from a population as complex and large as that of a nation is far from simple. Statisticians have developed a number of sampling strategies that allow researchers to collect representative samples from highly structured populations. Although these sampling schemes are not usually part of the beginning course, *stratified sampling*, *cluster sampling*, and *systematic sampling* are a few methods that you may have encountered. These all tend to produce representative samples, but statistics based on data collected through these schemes have formulas for standard errors that are different from those that are covered in a beginning statistics course.

Essential Understanding 5h
The size of the sample relative to the population size is not an important factor in determining the accuracy of estimates.

Implications of random selection and random assignment

Essential Understanding 4d. *Random selection and random assignment are different things, and the type and scope of conclusions that can be drawn from data depend on the role of random selection and random assignment in the study design.*

The adjective *random* is used in many different contexts in statistics, but two important ones are *random sampling* (selecting objects from a population by choosing them at random) and *random assignment* (assigning subjects to experimental groups at random). These terms name two very important and distinct concepts. Students often confuse these concepts, but it is important to keep them distinct because they have different consequences (Derry et al. 2000).

The chart in figure 1.24 summarizes the differences in the types of conclusions that can be drawn, depending on whether a study incorporates random selection and/or random assignment. If well designed, experiments that apply random assignment allow us to infer cause-and-effect relationships between variables by eliminating or at least minimizing the role of confounding factors. Studies based on random samples allow us to generalize conclusions to the larger population. Ideally, a study has both characteristics, but this is not always achievable.

	Random sampling	No random sampling
Random assignment	Can infer causality *and* can generalize from sample at hand to larger population	Can infer causality, but cannot generalize from sample at hand to larger population
No random assignment	Can generalize from sample at hand to larger population, but cannot infer causality	Cannot generalize from sample at hand to larger population *and* cannot infer causality

Fig. 1.24. Conclusions that can be drawn from data under study, depending on study design

Evaluating Estimators: Big Idea 5

Big Idea 5. *Evaluating an estimator involves considering bias, precision, and the sampling method.*

Estimators are used to approximate population parameters, but how can we assess how good an *estimator*, and hence, an estimate, is? For example, two commonly used statistics are the sample mean and the sample standard deviation, which estimate the population mean and the population standard deviation, respectively. But these are not the only measures of center and variability, so how can we evaluate and compare the quality of the various estimators?

The sampling distribution—the focus of Essential Understanding 2c—provides us with characteristics of sample statistics and can be used to evaluate how well a statistic serves as an estimator of a population parameter. For example, the center of the sampling distribution tells us what values the estimator will typically produce. Furthermore, if this center is close to the population value, we know that our estimator is accurate in its estimations. The spread of the sampling distribution helps us measure the *precision* of an estimation procedure. A statistic whose sampling distribution has great variability is one that produces very different estimates from one sample to the next and therefore has low precision.

Earlier, we highlighted the importance of collecting data properly. Recall that statistical methods involve using available, but usually incomplete, information to draw conclusions about a population or about the effect of experimental treatments on some response. Because such conclusions are based on incomplete information, an associated risk of error needs to be acknowledged and quantified. Characteristics of the data collection method—such as random selection and random assignment—make this possible. In addition to these characteristics, sample size must also be taken into account in quantifying the risk of error.

Some fundamental concepts of sampling are elusive. Many people struggle with the idea that a relatively small sample from a population can be used to estimate characteristics of the entire population accurately. In other words, they question how we can reasonably and confidently generalize from a sample to the population from which the sample was drawn. Another idea that many people mistakenly believe is that a larger population requires a larger sample.

Clarifying these ideas depends on understanding two factors—the sampling method and sample size—and the interplay between them. A well-designed sampling scheme can allow generalization to the population even if the sample is relatively small. In fact, a larger sample selected by using a poor design often produces worse results

Essential Understanding 2c
The sampling distribution of a sample statistic describes how the value of the statistic varies from sample to sample.

For an introductory discussion of simple random samples and sampling distributions, see *Developing Essential Understanding of Statistics for Teaching Mathematics in Grades 6–8* (Kader and Jacobbe 2013).

than a small sample selected by using a good sample design. Eight essential understandings support Big Idea 5 by elaborating on the relationships among estimators, precision, and sample size.

The quality of estimators

Essential Understanding 5a. Estimators are evaluated on the basis of their performance in repeated sampling.

A first course in statistics introduces students to just a handful of estimators. For example, students use the sample mean to estimate the population mean, which is a measure of the center of a population distribution. They use the sample standard deviation to estimate the population standard deviation, a measure of variability. But you are probably familiar with other measures of center, such as the median, and other measures of spread, such as the interquartile range. How can we evaluate and compare the quality of estimators?

One approach is to consider how an estimator was derived—its pedigree, if you will. Sometimes estimators are derived by using some mathematically based strategy (for example, in regression, least squares is used to estimate the slope and intercept). But at the end of the day, what matters is whether these estimators are any good. One reason that random sampling is so important is that it allows us to quantify the uncertainty in our estimation and therefore to evaluate the performance of statistics used as estimators.

Another, complementary approach is to consider how the estimator would behave if we were to take repeated samples from the population. In real life, we get only one sample, but the sampling distribution gives us a mechanism for asking what *would* happen if we could take a random sample again and again. Reflect 1.29 asks you to consider a situation in which several different estimators of a population characteristic are sensible.

Reflect 1.29

In World War II, American intelligence discovered that German tanks were stamped with consecutive serial numbers, beginning with the number 1. Thus, the serial number revealed some information about how many tanks had been produced. Suppose you are handed a list of 10 serial numbers of German tanks and are told that you can consider these as a random sample from the population of all the German tanks produced.

a. Think of two different ways of estimating the total number of German tanks.

b. Is it possible to decide, without knowing the true total number of tanks, which estimator is better?

c. What does "better" mean in this case?

The Big Ideas and Essential Understandings

A natural statistic to use in estimating the maximum value in a population is the maximum of a sample. The "strategy" employed here might be described as common sense. We might also think of using the sample mean plus three standard deviations, reasoning, as many students do, that in a symmetric distribution, it is rare to get a value larger than that sum, so that value might possibly be a good estimate for the maximum value in a population. There are many other estimator candidates, and a class of students might easily come up with a dozen or more.

If we hosted a contest to see which estimator is best, we might be tempted to say that the estimator that comes closest to the truth is the winner. But the value of the estimator is computed with data from one random sample, and the values of our candidate estimators will vary from sample to sample. We do not want to choose a winner just because it happens to be close to the population parameter in the case of one particular sample. The winner should be an estimator that was determined through an approach that works well in general. In other words, we want to reward sound methods, not ones that just happened to work well for one particular "lucky" sample.

Potential bias in estimators

Essential Understanding 5b. *Some estimators are biased.*

The following joke is popular among statisticians: Three statisticians go out duck hunting. A duck flies overhead. The first statistician shoots, but the bullet flies one meter above the duck. Almost simultaneously, the second statistician shoots, but this time the bullet flies one meter below the duck. The third statistician shouts, "We got 'im!"

The joke has meaning—and maybe is even funny—only if you understand a very important concept in statistics—bias. Bias measures how close, on average, the bullets come to the duck. Technically speaking, the bias of a statistic is the difference between the mean of the statistic's sampling distribution and the actual value of the parameter that this statistic is trying to estimate (for more detail about the sampling distribution, see our earlier discussion of Essential Understanding 2c). Unbiased estimators—that is, estimators that, on average, equal the value of the parameter—are usually desired. The hunting statisticians are happy because their shots are unbiased. One hunter shoots a meter above the duck (+1), the other a meter below (–1), and so, on average, they hit the duck. Reflect 1.30 asks you to consider bias in the situation where the sample maximum is used as an estimator of the maximum value in a population.

Essential Understanding 2c
The sampling distribution of a sample statistic describes how the value of the statistic varies from sample to sample.

> **Reflect 1.30**
>
> If the maximum of a sample of tank serial numbers is used to estimate the maximum value in the population of tanks, is it *unbiased*, *biased high* (tends to give values that are too great), or *biased low* (tends to give values that are too small)?

The maximum of a sample can never be greater than the maximum of a population. At most—and rarely—the maximum of the sample might equal the population maximum. For this reason, the maximum of a sample will typically be less than the population maximum, and this estimator is therefore *biased low*. It is important to note that this bias is a property that we can observe only through repeated sampling. Although in the tank example it is possible (although unlikely) that the estimator will exactly equal the population value, this is not the typical outcome.

When used to estimate the population mean, the sample mean is an example of an unbiased statistic. To illustrate this fact, consider the population consisting of Academic Performance Index (API) scores for 1915 California high schools reporting an API (not all schools reported this score). The mean of this population is 655. Figure 1.25 shows the results of a simulation study. Each point represents a sample mean API score based on a random sample of 10 high schools. The dot plot approximates the sampling distribution of the sample mean and is centered at 655. In general, the sampling distribution of the sample mean is centered at the actual population mean. Because the center of the sampling distribution is the same as the value of the population mean, the sample mean is an unbiased estimator.

When choosing between different estimators of the same parameter, statisticians often prefer the estimator with the least bias. For example, consider the estimator s^2 for the population variance:

$$s^2 = \frac{\sum(x - \bar{x})^2}{n-1}$$

You might have wondered why we divide by $n - 1$ (the sample size minus 1) rather than by n. The reason is that the estimator s^2 is unbiased. If we divide by n instead, the estimator can be shown to be biased by a factor of

$$\frac{n-1}{n}.$$

The preferred estimate of the population standard deviation is s (the square root of s^2), even though it is biased. It is preferred because the bias is small compared with that of other estimators, and the bias is less than if we had instead divided by n.

In many contexts, it is useful to think of bias as a measure of accuracy. An accurate duck hunter, for example, is one who tends

The Big Ideas and Essential Understandings

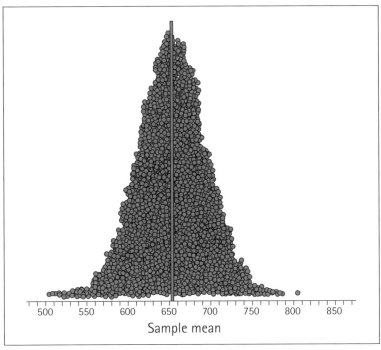

Fig. 1.25. An approximate sampling distribution of the sample mean

to hit the duck (although it might take several shots to do so). Remember that the bias measures a general tendency of an estimator across multiple samples. In real life, we usually have just one sample. We cannot use the results of a single sample to determine whether or not an estimator is biased. A value that is too large (or too small) may result simply because of random variation from sample to sample. Instead, information about bias must be based on theoretical considerations or on simulation studies.

Precision based on standard error and bias

Essential Understanding 5c. The standard error describes the precision of an estimator.

The variability of an estimator is measured by the standard deviation of the sampling distribution, called the **standard error**. The standard error measures the sample-to-sample variability of a statistic—a useful measure because it tells us how much we should expect the value of the estimator to vary from one sample to another. For example, the standard deviation of the 10,000 sample mean API scores in figure 1.25 is 38.3. If we were to take a random sample of 10 schools, we would expect the mean of this sample to be around 655 (because the sample mean is an unbiased estimator of the population mean), but observed sample means will vary, typically falling about 38.3 points from 655.

> **Essential Understanding 2d** Simulation can be used to approximate sampling distributions.

For many estimators, the standard error is estimated on the basis of theoretical considerations. For example, we saw in our discussion of Essential Understanding 2d that the standard deviation of the sampling distribution for the sample mean (called the standard error of the sample mean) is approximately

$$SE_{\bar{x}} = \frac{s}{\sqrt{n}},$$

where s is the standard deviation of a random sample of size n from the population. We can use this formula to calculate the approximate standard error for estimating the population mean API score by using the sample mean. Suppose that a random sample of 10 high school API scores is as follows:

655 840 686 618 644 795 433 739 514 576

The sample mean for these ten scores is $\bar{x} = 650.0$, and the sample standard deviation is $s = 124.047$. The approximate standard error is therefore

$$SE_{\bar{x}} = \frac{s}{\sqrt{n}} = \frac{124.047}{\sqrt{10}} = 39.2.$$

This result is fairly close to the value of the standard deviation obtained from the simulation.

The formula for the standard error of the sample mean is informative because it tells us that if we take a larger sample, we will have less variability in the values of our estimator. The standard error can be thought of as a measure of the precision of an estimator. The smaller the standard error, the more precise the estimator. In other words, the smaller the standard error of the sample mean, the less the value of the sample mean will vary from sample to sample, and the more tightly the values of the sample means will cluster around the actual value of the population mean. The formula

$$SE_{\bar{x}} = \frac{s}{\sqrt{n}}$$

shows us that in the case of the sample mean, the larger the sample size n, the smaller the standard error, and thus the more precise our estimator.

The standard error is not as easily determined for many estimators as it is for the sample mean. The standard error of the sample median, for example, depends on the shape of the population distribution in a complex way and is not easy to estimate by applying theory.

In most contexts, an estimator with little or no bias and small standard error is preferred. This is one reason why the sample mean is so widely used. It is always unbiased (if based on a random sample), and the standard error can be made arbitrarily small by using a large sample size. Reflect 1.31 asks you to revisit the German tank example and examine the approximate sampling distributions

The Big Ideas and Essential Understandings

in figure 1.26 to consider how bias and precision might be used to compare estimators.

Reflect 1.31

Figure 1.26 shows approximate sampling distributions based on a population of German tanks in which the maximum number was 562. For each of the 1000 simulated sampling repetitions, a random sample of 7 "tanks" was selected. The graphs show approximate sampling distributions for three different estimators of the maximum number of German tanks. The true value of the total number of tanks in this example is 562 and is indicated with a vertical line. Each point represents a different random sample of seven tanks.

Which estimator do you think is best? Why?

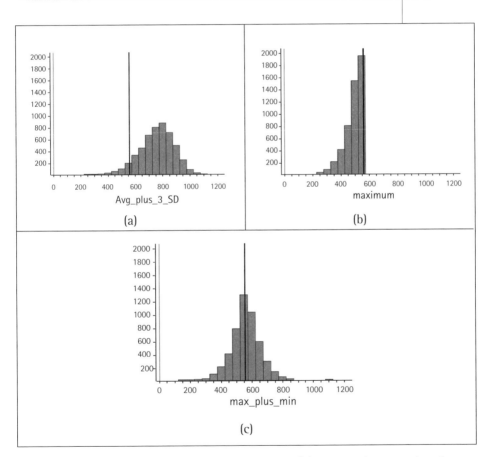

Fig. 1.26. Approximate sampling distributions for (a) the sample mean plus three standard deviations, (b) the maximum of the sample, and (c) the maximum of the sample plus the minimum of the sample

When evaluating an estimator, we usually take both the bias and the standard error into account. An estimator with small bias and high precision is usually preferred. Using the mean of the sample plus three standard deviations to estimate the population maximum typically produces estimates that are too high, as we can see from figure 1.26a, in which the mean of the sampling distribution is about 755, a bias of 193 points. For this estimator, the approximate standard error, calculated by finding the standard deviation of the simulated sample means used to produce the figure, is roughly 119 points. The sample maximum, as we predicted earlier, is biased low by approximately (500 − 562), or −62, points (see fig. 1.26b).

The third estimator (fig. 1.26c), the sample maximum plus the sample minimum, is in fact unbiased. Its standard error is roughly 92 points. Although it does not have the smallest standard error of the three, the fact that it is unbiased and has a relatively small standard error means that in many contexts, it would be the preferred estimator of these three.

We might have expected that this estimator would be unbiased. We know that the sample maximum is biased low. How low? We might expect that it is about as far from the population maximum as the sample minimum is from the population minimum. Therefore, adding the sample minimum to the sample maximum "makes up for" the low bias resulting from using only the sample maximum. (Incidentally, in the context of the German tanks, there is a "best" estimator, which is unbiased and has the highest achievable precision. See the German Tank Problem on the Internet—for example, at http://en.wikipedia.org/wiki/German_tank_problem.)

Confidence intervals as estimators

Essential Understanding 5d. Confidence intervals are estimators that convey information about precision.

Confidence intervals use the standard error, a measure of the variability of an estimator, to provide a range of "plausible" values for the population parameter. The width of the interval is determined in part by the standard error, but also by a confidence level that is associated with the interval. Because a confidence interval estimate is computed from sample data, when we create the interval and state that we think that the interval includes the actual value of the population parameter that we are estimating, there is a chance that we are wrong. The **confidence level** quantifies this chance of being incorrect. Usual choices for confidence levels are 90%, 95% and 99%. A 90% confidence interval is one that is computed by using a method that will be "right" 90% of the time and "wrong" 10% of the time.

For example, to make figure 1.27 (generated by use of an applet at www.rossmanchance.com), we took 100 random samples, each of size 10, from the population of high school APIs. For each

The Big Ideas and Essential Understandings

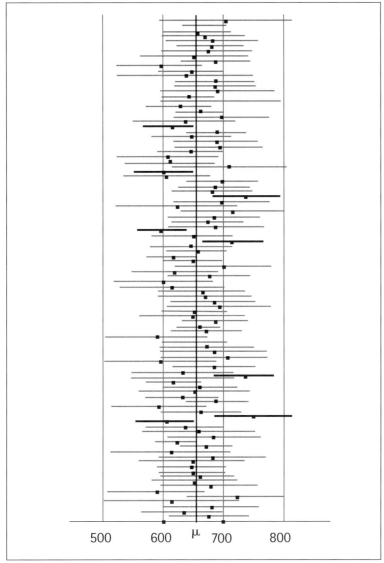

Fig. 1.27. One hundred 90% confidence intervals, each based on a random sample of 10 API scores drawn from a large population

random sample, we calculated a 90% confidence interval for the mean API score. Theory tells us, and simulations confirm, that about 90% of these intervals will contain the true value of 655. We expect that about 10% of the intervals will not include the population value (655), although in this particular simulation, only 8 did not include 655.

Confidence interval estimates are computed by forming the interval

statistic ± margin of error.

Informally, the **margin of error** can be thought of as the maximum likely estimation error—that is, the most that we think it is likely that the value of our sample statistic will differ from the actual value of the population parameter. Formally, the margin of error is defined as

$$K \cdot (\text{standard error of the statistic}),$$

where K is a number that depends on the confidence level and the sampling distribution of the statistic.

Two things are noteworthy about the margin of error. First, the number K is a function of the confidence level. The higher the confidence level, the greater the value of K and the larger the margin of error. This makes sense because a higher confidence level corresponds to a method that is "right" more often—in other words, the interval includes the actual population value more often. This is accomplished by widening the interval. Second, recall that the standard error of a statistic is a function of sample size, with larger sample sizes being associated with smaller standard errors. This means that, for a fixed confidence level, a smaller margin of error can be achieved by increasing the sample size, resulting in a narrower (more precise) confidence interval.

Although the calculation of confidence intervals is beyond what can be covered in a book of this size, conceptual understanding of confidence interval estimates and of confidence levels are important. The width of a confidence interval conveys information about the precision of the estimate. For example, suppose that (4.2, 23.6) is a 95% confidence interval for the mean number of hours that high school students spend online in a typical week. This interval is very wide, and it tells us that although we believe that the actual mean is between 4.2 hours and 23.6 hours, we do not really have a precise estimate of the actual mean time. On the other hand, if the 95% confidence interval had been (17.4, 18.1), we would have a much more informative estimate of the population mean. Reflect 1.32 asks you to consider why we cannot use a 100% confidence level.

Reflect 1.32

Why settle for 95% confidence level? Why not use a 100% confidence level?

A confidence level of 100% can be achieved trivially by choosing an infinitely wide interval: $(-\infty, +\infty)$. Although this interval delivers the advertised 100% coverage probability, it falls short in the "usefulness" category since it tells us nothing about the value of the population parameter that we wanted to estimate. Narrower intervals provide more information but are accompanied by smaller levels of confidence. An interval with a confidence level of 10% might convey a very precise location for the population parameter, but because this method fails for 90% of all random samples, this precise

The Big Ideas and Essential Understandings

estimate is probably not useful. A confidence level of 95% strikes a nice balance, in many contexts, between precision and confidence.

One idea that students struggle with when working with confidence intervals is the notion that the confidence level is a measure of the quality of the *method*, and not a measure of the quality of an *individual interval*. The blurring of this distinction is what irritates statisticians in hearing someone state that a confidence interval of, say, (570, 758) "has a 95% probability of containing the true population mean API score of California high schools." This statement attaches a probability to the particular interval—in this example, (570, 758)—when, in fact, the probability should be attached to the process of collecting and using a random sample to compute confidence intervals from this population.

To help understand how to interpret a confidence level, consider a metaphor. Think of a car factory that has a quality rating of 95%—that is, 95% of the cars that it produces are good cars. This does not mean that your car, purchased from this factory, will be good 95% of the time. Reflect 1.33 asks you to think about the meaning of 95% confidence.

> **Reflect 1.33**
>
> On the basis of a random sample of schools, the interval (570, 758) is a 95% confidence interval for the mean API of California high schools. Why is it incorrect to say, "This interval is good 95% of the time"?

In the mythical car factory that we imagined, 5% of the cars are lemons. Once you purchase a car, it is either a lemon or a good car. A probability is no longer associated with this circumstance, since you will know when you drive the car. (Let's assume, for the sake of simplicity, that the "bad" cars are bad in a way that is immediately obvious.) To say, "There's a 95% chance that my new car is good" implies that there is variability in your car. Because the car does not suddenly change into another car, the only way there could be variability would be if the car changed over time. Although it might be true that your car will sometimes need repair, it is not true that 5% of the time your car is a car with a manufacturing flaw, and the other 95% of the time the flaw disappears.

Confidence intervals work in the same way. They are produced in a "factory" and are based on an infinite supply of random samples of a fixed size. We get to "purchase" only one confidence interval, and our interval is either good or bad. Because the interval does not vary once we've taken the random sample, there is no variability to which we can "attach" a probability. This is why the "95%" applies to the method of producing confidence intervals. We can say that there is a 95% chance that the method will produce a good interval, but we cannot assign a probability to any particular interval produced by the method.

Precision depends on sampling method and sample size

Essential Understanding 5e. *The precision of estimators depends both on the way in which the sample was selected and on the size of the sample.*

Random selection of individuals in a population tends to produce samples that are representative of the population. Many different types of random selection methods, or *sampling schemes*, can be employed. Simple random sampling and stratified random sampling are just two of many possible schemes. In simple random sampling, every sample of size n from a population has the same chance of being selected. Simple random sampling is rarely used in large-scale studies, because other sampling methods may be easier to implement or may enable more precise estimates. However, simple random sampling does form a foundation for many of the other sampling methods.

Stratified random sampling incorporates the use of supplemental information to allow researchers to select a sample that can result in a more precise estimate of a population parameter. By definition, a **stratified random sample** divides the population into mutually exclusive subgroups, called **strata**, from which simple random samples are taken. The strata are created in such a way that the individuals within each stratum are as alike as possible with respect to the response variable of interest. In practice, strata are based on a characteristic believed to be associated with the response variable. For example, strata might be created on the basis of income if researchers believe that people with similar incomes might respond similarly to a survey question. Reflect 1.34 asks you to consider possible ways of creating meaningful strata.

Reflect 1.34

How might you create strata if you want to design a plan for selecting a sample of high school students to use in learning about the number of text messages that they send each month?

Consider, for example, the study of high school students' text messaging previously discussed. Suppose that we want to determine the mean number of text messages sent per student. The number might be very different for male students and female students. Or the number of text messages sent might differ by grade level. Simple random sampling would not take this additional information

The Big Ideas and Essential Understandings

into consideration. By using stratified random sampling, however, we could divide the students by gender or by grade level and then sample within each gender or grade. Note that if we are right that males and females have different texting habits, the advantage of our stratified sampling scheme is that each stratum is more homogeneous with respect to the variable of interest—the number of text messages sent—than would be the case for the entire population of students. Stratifying on a variable that is unrelated to the response variable is no more effective than simple random sampling. By contrast, successful stratification, when compared with simple random sampling with the same overall total sample size, results in smaller sample-to-sample variability, and hence, more precise estimates.

Sample size also plays a role in the precision of estimates. Figure 1.28 shows approximate sampling distributions, each based on 1000 sample means, for samples of size 10, 20, 40, and 80. Reflect 1.35 asks you to investigate this figure to review the effect of sample size on sample-to-sample variability.

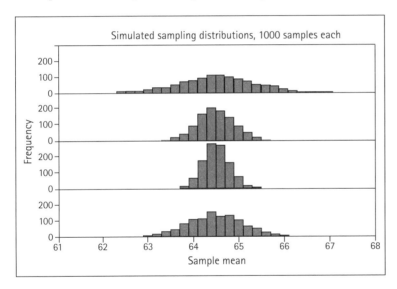

Fig. 1.28. Approximate sampling distributions based on 1000 sample means for samples of size n = 10, 20, 40, or 80

Reflect 1.35

Which of the approximate sampling distributions for the sample mean shown in figure 1.28 corresponds to samples of size 10? Which corresponds to samples of size 20? Samples of size 40? How do you know?

As the sample size increases, the standard deviation of the sampling distribution of the sample mean decreases. Therefore, the approximate sampling distributions in figure 1.28, from top to bottom, correspond to samples of size 10, 40, 80, and 20. Note that the smallest sample size has the largest standard deviation of sample means, and the largest sample size has the smallest standard deviation of sample means. As such, the largest sample size ($n = 80$, in this case) will provide the most precise estimate, and the smallest sample size ($n = 10$) the least precise estimate, of the population mean. We want to make our estimates as precise as possible. To achieve high precision, we want to take as large a sample as is reasonable, using a well-designed sampling method.

Sample size

Essential Understanding 5f. *If the sampling method is good, a larger sample is always more useful than a smaller sample.*

As we discussed previously, larger samples provide estimates that are more precise than smaller samples. Why, then, do introductory statistics textbooks sometimes indicate that for the use of a particular method to test a claim or estimate a population parameter, samples should be no larger than 10% of the population size? Would it not be better to take a larger sample to test a claim or make an estimate?

In fact, it *would* be better to take a larger sample rather than a smaller sample. Why then the apparent contradiction with the recommendation that samples be less than 10% of the population? The answer has to do with how the formulas for test statistics (when testing a claim) or confidence intervals (when making estimates) are derived, and the assumptions that underlie the derivations.

Consider the usual formula for the standard error of $\bar{x}$ used when testing a claim about μ or constructing a confidence interval for μ. The formula takes the form

$$SE_{\bar{x}} = \frac{s}{\sqrt{n}},$$

where s is the sample standard deviation, and n is the sample size. This formula is derived under the assumption that the n individuals in the sample are selected independently of one another. If the sampling is done with replacement, or if the population is infinitely large, then this is the appropriate formula for the standard error. However, sampling is usually done *without* replacement and from a finite population. Thus, when an individual is selected from the population, the makeup of the population remaining has changed. The result of sampling without replacement is that the selections are not truly independent of one another.

When researchers sample without replacement, they need to use a somewhat modified version of the expression for the standard error. The modification incorporates what is called the *finite population correction factor*, which takes into account the fraction of the population that they are not sampling. Specifically, the finite population correction multiplies the formula for standard error by the factor of $1 - \frac{n}{N}$.

In this multiplier, n is the sample size, and N is the population size, so $\frac{n}{N}$ is the proportion of the population included in the sample. In introductory statistics courses, the assumption is that if $\frac{n}{N}$ is equal to or smaller than .1, or 10%, then the finite population correction is negligible for practical purposes, and researchers can disregard the impact of the lack of independent selection. However, as $\frac{n}{N}$ gets larger, the finite population correction takes on greater significance, and researchers therefore should include it. The chart in figure 1.29 shows values for n, the sample size (between 10 and 1000); the proportion sampled, $\frac{n}{N}$ (assuming a population size of $N = 1000$); and the standard error, computed both without the finite population correction

$$\frac{s}{\sqrt{n}}$$

and with the finite population correction

$$\frac{s}{\sqrt{n}}\left(1 - \frac{n}{N}\right),$$

assuming $s = 10$. A fifth column in the chart shows the percentage difference between the two computations of the standard error. We will consider this column in discussing Reflect 1.36, which explores advantages of incorporating the finite population correction factor in the computation of the standard error.

n	Proportion sampled, $\frac{n}{N}$	$\frac{s}{\sqrt{n}}$	$\frac{s}{\sqrt{n}}\left(1-\frac{n}{N}\right)$	Percentage difference
10	0.010	3.162	3.131	1.01
20	0.020	2.236	2.191	2.04
40	0.040	1.581	1.518	4.17
100	0.100	1.000	0.900	11.11
200	0.200	0.707	0.566	25.00
500	0.500	0.447	0.224	100.00
510	0.510	0.443	0.217	104.08
700	0.700	0.378	0.113	233.33
800	0.800	0.354	0.071	400.00
1000	1.000	0.316	0.000	

Fig. 1.29. For selected values of n (sample size) in a population $N = 1000$, the proportion sampled $\left(\frac{n}{N}\right)$; the standard error, computed both without the finite population correction and with it, assuming $s = 10$; and the percentage difference between the two estimates of the standard error

Reflect 1.36

Review the information in figure 1.29. What advantages do you see in using the formula for the standard error that incorporates the finite population correction factor? When does using the finite population correction seem not to matter?

Notice that as $\frac{n}{N}$ gets larger, the standard error decreases, until we have sampled every individual in the population (i.e., taken a census), at which point we have no error in our estimate of the population mean (the standard error incorporating the finite population correction factor is zero). Although the standard error decreases toward zero as the sample size increases, the percentage difference between the standard errors computed both with and without the finite population correction increases. Note also that $\frac{s}{\sqrt{n}}$ always overestimates the standard error compared with $\frac{s}{\sqrt{n}}$ multiplied by the finite population correction factor, so the $\frac{s}{\sqrt{n}}$ form of the standard error gives a more conservative estimate. Once the proportion sampled exceeds about 10% (that is, $\frac{n}{N} > .1$), the percentage difference between the two computations of the standard error begins to become too large to be ignored. This is part of the justification for stating in introductory statistics textbooks that the sample should be no larger than 10% of the population. In fact, a larger sample is better than a smaller sample, but the finite population correction should be considered in computing the standard error when the fraction of the population sampled exceeds approximately 10% of the population. A smaller standard error is certainly more desirable since it indicates increased precision.

Another aspect of the effect of increasing the sample size must be considered. The information in figure 1.29 shows that the standard error decreases as the sample size n increases. What is less obvious is that the standard error decreases significantly for an increase in a small sample size, but the decrease in standard error becomes smaller for increases in already large samples. Figure 1.30 illustrates this fact for the information in figure 1.29. In the plot, the variable on the horizontal axis is the sample size, and the variable on the vertical axis is the standard error, computed both with and without the finite population correction. Reflect 1.37 asks you to examine the effect of increasing the sample size n on the standard error.

Reflect 1.37

Referring to the plot in figure 1.30, describe the relationship between the standard error (both with and without the finite population correction) and the sample size n. What are the implications of this observed relationship?

The Big Ideas and Essential Understandings

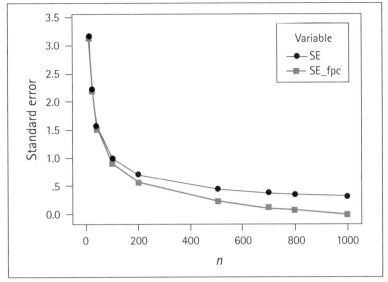

Fig. 1.30. A plot of standard error vs. *n*, the sample size, assuming a population of size 1000 and standard deviation *s* = 10

Notice that an increase in sample size from 10 to 20 (an increase in sample size of 10), for example, corresponds to a decrease in standard error of approximately 30%, whereas an increase in sample size from 500 to 510 (also an increase in sample size of 10) corresponds to a decrease in standard error of approximately 3%.

In general, when we are not considering the finite population correction, we must quadruple the sample size to halve the standard error. Therefore, increasing the sample size from 1 to 4 will halve the standard error, and, likewise, increasing the sample size from 200 to 800 will also halve the standard error. As we can see, increasing the sample size results in both a decrease in the standard error and a corresponding increase in precision. However, there is a point of diminishing returns, at which the additional cost associated with increasing sample size may not be worth the gain in precision. Note that this relationship also holds for the standard error of sample proportions.

Good methods trump large samples

Essential Understanding 5g. *A small sample selected by using a good method can yield better results than a large sample selected by using a poor method.*

History buffs and statistics buffs alike may have seen the famous photograph of Harry Truman holding up a copy of the *Chicago Daily Tribune* on November 3, 1948–the day after Truman defeated Thomas Dewey in an upset election for the U.S. presidency. "Dewey Defeats Truman," the headline declared, and the *Tribune* ran it on the basis of the results of several polls predicting that Dewey would defeat Truman.

Earlier, in the 1936 U.S. presidential election, the *Literary Digest* had erroneously predicted that Republican Alf Landon would win 55% of the popular vote, and President Roosevelt would win just 41%. Roosevelt actually received 61% as compared with Landon's 37% of the popular vote. What could have gone so wrong? After all, the *Literary Digest* had a streak of predicting the winners of U.S. presidential elections from 1912 through 1932. Reflect 1.38 asks you to consider what could have contributed to the incorrect predictions in the 1936 and 1948 elections.

Reflect 1.38

What are some possible reasons why the polls were wrong in the 1936 and 1948 U.S. presidential elections?

Problems like these can arise as a result of a poor data collection plan. In the case of the 1936 election, the *Literary Digest* survey sent out 10 million surveys, but only 2.3 million were returned. Although the number of surveys returned was quite large, the response rate was only 23%. In addition, the 10 million U.S. households that the *Digest* selected for participation in the survey were chosen from telephone directories and vehicle registration records. At the time, only the more affluent households had telephones or motor vehicles. As a result, the population as a whole was not adequately represented in the sample, and respondents who were more affluent might have been more likely to vote Republican. An entire segment of the population of U.S. households—namely, those without telephones or vehicles—was excluded from the sample.

Although the *Literary Digest* had a huge sample, its prediction was wrong. By contrast, in the same election, George Gallup used a sample of only 50,000 and predicted that Roosevelt would beat Landon. Although Gallup's sample was considerably smaller than that of the *Literary Digest*, it was more representative of the voting population.

In summary, a large sample size does not guarantee the accuracy of estimates made from that sample. In fact, statistician Sharon Lohr (2009) asserts that a large unrepresentative sample may do more damage than a small one because many people think that large samples are always better than small ones. She notes that the design of the sampling plan is far more important than the absolute size of the sample.

The Big Ideas and Essential Understandings

Small samples despite large populations

Essential Understanding 5h. The size of the sample relative to the population size is not an important factor in determining the accuracy of estimates.

One final misconception about sampling needs to be addressed—the idea that larger populations require larger sample sizes. Consider, for example, a survey conducted in the states of California, which has the largest population of all the U.S. states, with 37,253,956 citizens, and Wyoming, which has the smallest population, with 563,626 citizens (2010 U.S. Census, www.census.gov). For a survey conducted in these two states, many people would think that California requires a larger sample size than Wyoming to achieve the same level of precision. They would suppose that since the population of California is approximately 66 times that of Wyoming, it is obvious that California requires a larger sample. In fact, to achieve essentially the same level of precision with both states, and hence have the same risk of error associated with their corresponding estimates, a survey conducted in these two states could use samples of the same size, as long as the sampling method is good.

How can this be? Recall that precision is used to assess the quality of estimates or claims about population characteristics, and it is quantified by the standard error. In examining Essential Understanding 5f, we saw that when the sample size is no greater than 10% of the population size, then the approximate standard error of the sample mean is

$$\frac{s}{\sqrt{n}}$$

Notice that the population size, N, does not appear in that expression, so clearly population size does not play a significant role in computing the approximate standard error for samples that are small relative to the population size. We also saw that the formula for the standard error for sample means that incorporates the finite population correction factor has the form

$$\frac{s}{\sqrt{n}}\left(1 - \frac{n}{N}\right).$$

This is the form of the standard error needed when the sample size is at least 10% of the population size; note that the population size N is part of this formula.

Figure 1.31 illustrates the relationship between standard error and population size for fixed sample sizes. It plots standard error, computed by using the finite population correction, versus population size, for samples of size n = 20, 40, 80, and 100, assuming a fixed population standard deviation of σ = 20. Reflect 1.39 asks you to consider the effect of population size on the standard error for samples of various sizes.

Essential Understanding 5f
If the sampling method is good, a larger sample is always more useful than a smaller sample.

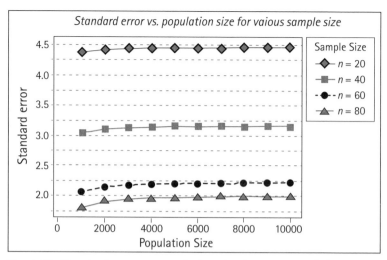

Fig. 1.31. Standard error vs. population size for samples of size 20, 40, 80, and 100, all with a population standard deviation σ = 20

Reflect 1.39

What key observation can you make regarding the standard error for fixed sample size as a function of population size, as illustrated in figure 1.31?

First note that, as we have previously discussed, smaller samples have larger standard errors for a fixed population size, an idea that is apparent in figure 1.31. Also note that as the population size, N, increases, for a fixed sample size, n, and standard deviation, σ, the standard error levels off and does not change significantly once the population reaches a certain size.

Again consider the California and Wyoming scenario. Assuming that a simple random sampling method has been used to take a sample of size $n = 1500$, the standard error that includes the finite population correction factor can be computed for each state (assuming a population standard deviation $\sigma = 20$). For Wyoming, the standard error of the sampling distribution of sample means is

$$\left(\frac{20}{\sqrt{1,500}}\right)\left(1 - \frac{1,500}{563,626}\right) = 0.515,$$

and for California, the standard error of the sampling distribution of sample means is

$$\left(\frac{20}{\sqrt{1,500}}\right)\left(1 - \frac{1,500}{37,253,956}\right) = 0.516.$$

Notice that these values differ by only 0.001, or 0.26%. Other states with populations closer in size will differ by an even smaller percentage for $n = 1500$ and equal standard deviations. For most practical purposes, this difference is inconsequential when making estimates or testing claims.

Misconceptions abound when it comes to sampling. Fundamentally, the interplay between sampling scheme and sample size are critical to understanding how we can safely and accurately make estimates or test claims about a population. Smaller samples taken by using good sampling methods can often lead to accurate results. In addition, the population size has an essentially negligible impact on the precision of estimates.

Conclusion

The big ideas that we have examined in this chapter focus on the role of variability and how variability plays out in statistical inference. Big Ideas 1 and 2 focus on the importance of considering variability in modeling real processes and of using distributions (population, sample, and sampling) to describe variability in data and in the values of sample statistics. Big Idea 3 captures some of the concepts that provide the foundation for hypothesis testing, as well as the role that variability plays in permitting us to assess both the strength of the conclusions that can be drawn in the face of uncertainty as well as the risk of possible errors in the conclusions. Big Idea 4 centers on data collection, and embedded in it is an understanding that the method used to collect data (through designed experiments or statistical sampling) determines the type of conclusions that can be drawn and how these conclusions might be generalized. Big Idea 5 extends this idea and captures some common misconceptions in sampling, including those related to sample and population sizes.

Statistical literacy and the ability to evaluate conclusions critically are essential to being an educated citizen in today's society. Newspapers, magazines, news reports, and politicians all quote results of studies—both experiments and surveys. The key to statistical literacy is being able to evaluate the results of such studies objectively and analytically. This capacity includes understanding the importance of data collection and its implications for the specific conclusions that can reasonably be drawn from the data. In addition, consumers of statistics must recognize the risk of error that is inherently associated with using statistical methods to draw conclusions from data.

Big Idea 1

Data consist of structure and variability.

Big Idea 2

Distributions describe variability.

Big Idea 3

Hypothesis tests answer the question, "Do I think that this could have happened by chance?"

Big Idea 4

The way in which data are collected matters.

Big Idea 5

Evaluating an estimator involves considering bias, precision, and the sampling method.

Chapter 2

Connections: Looking Back and Ahead in Learning

Now is an exciting time to be teaching statistics, but with this excitement come new challenges. In the next few years, most states' adoption of the Common Core State Standards for Mathematics (CCSSM) (Common Core State Standards Initiative [CCSSI] 2010) will lead to a deep integration of statistics into the mathematics curriculum. It is no exaggeration to say that by 2014, almost every teacher of algebra 2 will also be a statistics teacher. This dramatically changing landscape is exciting because statistics is being integrated into our culture and society as never before, and every student needs to achieve some level of statistical literacy, regardless of his or her future academic choices. Today, nearly every college major, whether "technical" or not, requires at least one statistics class.

Differences between Statistics and Mathematics

The primary challenge that mathematics teachers face with respect to the shift to a curriculum aligned with CCSSM is that statistics differs from mathematics in fundamental ways. Mathematics is built on a foundation of deduction, proof, and truth, whereas statistics is based on inference, data analysis, and uncertainty. Currently, school curricula develop mathematical reasoning slowly over the entire first twelve years of a student's academic life; by contrast, students in many schools are expected to learn statistical reasoning in a single one-term or one-year course, if at all. The adoption of CCSSM means that teachers will now have the opportunity—and the responsibility—to develop students' abilities to reason inferentially with data.

The big ideas in this Essential Understanding volume highlight conceptual milestones in the development of statistical reasoning skills. While we have chosen milestones that are particularly important for teachers' understanding, they are also important for student understanding, and this chapter emphasizes stages in students' development of the five big ideas that we have chosen to highlight. The pathway to developing understanding is perhaps less clear than that for the development of mathematical understanding, in part because the topic is less familiar to mathematics teachers, but also because the development of statistical reasoning does not follow a trajectory of logic-driven growth from basic to complex concepts. Instead, statistical understanding begins experientially and gradually adds layers of conceptual complexity.

Informal to Semiformal to Formal Reasoning

NCTM's *Principles and Standards for School Mathematics* (2000) sketched out a pathway for developing statistical reasoning and inference. This progression was developed more fully in the pre-K–12 curricular framework presented in the American Statistical Association's *Guidelines for Assessment and Instruction in Statistics Education* (GAISE Framework) (Franklin et al. 2007)—and more recently in the Common Core State Standards for Mathematics (CCSSI 2010). This pathway can be seen as a journey along a spectrum from informal to semiformal and finally to formal statistical reasoning. Seen in this light, a high school curriculum consistent with CCSSM and with the big ideas addressed in this Essential Understanding volume would cover, for the most part, semiformal statistical reasoning.

At the earliest levels, informal statistical reasoning occurs when students apply everyday knowledge to the making of predictions, judgments, and claims about a population, based on a sample (Makar and Rubin 2009; Zieffler et al. 2008). As students progress, they learn to integrate statistical and mathematical concepts into this inferential reasoning. An important component of this intermediate, or semiformal, stage is developing ways of thinking about how to evaluate inferential procedures. At the formal level, students employ well-established statistical methods, articulate arguments supporting the use of these methods for the problem at hand, and interpret and communicate results in language appropriate for the intended audience.

The progress from informal to semiformal and then to formal reasoning is also elaborated in the GAISE Framework, which describes students' progress as taking them from understanding variability within a sample, to understanding that this variability might reflect the variability of the population from which the sample was selected, and finally, at the most advanced level, to understanding how to draw conclusions about populations on the basis of samples.

Reasoning Related to Each Big Idea

In a curriculum aligned with CCSSM, students' understanding of each of the five big ideas related to statistical reasoning progresses from an informal to a semiformal stage in the K–12 years. In this section, we highlight this progression from informal to semiformal understanding for each of the big ideas, and we describe how formal understanding is developed in subsequent course work.

Only rarely is the development of formal statistical understanding a goal at the high school level. The curriculum for Advanced Placement (AP) Statistics and college-level introductory statistics courses formalize many of the concepts embedded in the big ideas of chapter 1. With formalism comes the danger of rote learning and mindless application of procedures (Diaconis 1981).

Students need a firm understanding of the big ideas of chapter 1 if they are to understand the more formal approach and apply statistical procedures mindfully, with an awareness of their strengths and limitations. The description of the development of statistical thinking that follows assumes a school curriculum that reflects recommendations for statistics that are similar to those in the Common Core State Standards for Mathematics. Specific examples of activities that develop students' understanding in a manner that matches the progression that we describe can be found in the appendix of the GAISE Framework for pre-K–grade 12 (Franklin et al. 2007).

The growth of Big Idea 1

Big Idea 1 acknowledges that variability is inherent in real-life processes. This big idea captures the notion that "pure" mathematical processes represent trends, and distributions represent variability. In a curriculum aligned with CCSSM, students first encounter the concept of variability informally for one variable in grade 6, and they use the mean absolute deviation (MAD) to quantify variability about the sample mean.

For example, a classroom activity described in the GAISE appendix asks students to examine the variability in the length of first names of students in the class. Once students have determined the mean, they can find how much their own name's length differs from the mean. The average of the absolute values for all students in the class provides an informal measure of the variability.

In grade 8, students extend these ideas to two variables. They might, for example, examine data that show that height and arm length are variable for students in their class, yet also vary together, since taller students tend to have longer arms.

Big Idea 1

Data consist of structure and variability.

In grades 9–12, students explicitly acknowledge the concept of variability about a central value and begin to analyze how well statistical models fit actual data. They encounter more "formal" measures of variability, such as the standard deviation. They can see the standard deviation as an extension of the concept of mean absolute deviation—the MAD—encountered in earlier grades, and investigation might show them, for example, that most students will have an arm length within three standard deviations of the mean.

Students who pursue statistics beyond high school will use the increasingly formal understanding that they develop to specify models with mathematical notation and quantify goodness of fit by using measures such as r-squared and by quantifying the variability of residuals. They will use this more sophisticated understanding to develop more complex models, such as multiple regression models.

 Big Idea 2

Distributions describe variability.

The growth of Big Idea 2

CCSSM calls for introducing students to sample distributions in grades 6–8. Students first summarize and describe discrete distributions in grade 6, and they learn through tactile simulation to see them as descriptions of data generation processes that might come, say, from flipping a coin. At that time, students also learn that samples of data can be described with a distribution and that distributions can be summarized in terms of their shape, center, and variability. Numerical measures of center and variability are also introduced in the middle grades, although some of these, such as the mean absolute deviation, are intended to increase conceptual understanding and are not generally used as formal statistical measures.

The sampling distribution is a new concept that students encounter in high school, but their understanding is based on earlier, informal work in grade 7, where students have the experience of taking multiple samples to "gauge how far off the estimate might be" (CCSSI 2010, CCSSM, Grade 7, Statistics and Probability, p. 50). Students might, for example, explore how estimates of the probability of a coin landing on heads are not the same for every sample of coin flips.

In high school, ideas about types of distribution are united, and the relationships among population distribution, sample distribution, and sampling distributions are made explicit. The concept of a distribution is developed even further. For instance, students learn about continuous probability distributions, such as the normal distribution.

In subsequent study, students will continue to formalize the sampling distribution concept, enabling them to evaluate probabilities (for example, the probability that a sample mean will be within 1.5 standard deviations of the population mean) by using sampling distributions for a variety of statistics.

Connections: Looking Back and Ahead in Learning

The growth of Big Idea 3

In grade 7, students compare two samples and draw informal inferences about two populations. For instance, they might observe that the average weight of a random sample of boys' backpacks is greater than the average weight of a random sample of girls' backpacks, decide that this difference is large, and so conclude that in the population, boys tend to carry heavier backpacks than girls. The informal decision that a difference is "large" is based on the experience of taking random samples from simulated populations.

In high school, this reasoning process becomes semiformal through the introduction of a reasoning process that probes whether an observed difference could be explained simply by chance when one of two competing hypotheses is true. Students learn to state and informally test a null hypothesis that boys' and girls' backpacks tend to weigh the same, for instance. Students also begin to quantify the rate at which errors may result from this reasoning process. The idea that we can sometimes rule out chance as an explanation for observed differences becomes explicit, and this reasoning is linked to the data collection process.

In the future, students who pursue statistics will formalize these ideas in a collection of hypothesis test procedures designed to answer specific questions in a variety of settings. For instance, students might examine data to determine whether a particular diet did indeed lead to long-term weight loss, or whether preference for a popular fast-food restaurant is associated with a particular political affiliation.

Big Idea 3

Hypothesis tests answer the question, "Do I think that this could have happened by chance?"

The growth of Big Idea 4

In grade 7, students learn that the quality of an answer to a statistical question depends on the method of data collection. Their experiences with different types of data and questions related to data make them aware of the existence of surveys and experiments, the need for collecting data to answer questions, and the distinction between statistical questions (questions that involve variability) and nonstatistical questions. Seventh graders' understanding of these ideas then prepares them to learn about random sampling from a population.

In high school, students extend this knowledge as they learn to draw distinctions among the different data collection methods of surveys, observational studies, and controlled experiments. Students can read descriptions of studies reported in the media and determine the study design used, and evaluate whether the reported conclusions are appropriate for the given design. They see the role that controlled experiments and random assignment play in providing evidence for causal inference. An understanding of data collection and its implications for inference is a critical component of statistical literacy.

Big Idea 4

The way in which data are collected matters.

Students who take courses beyond high school will deepen and apply a more formal understanding of data collection. As they examine different methods for conducting surveys and experiments, they will see how these methods can be used to improve the precision of estimators and reduce the risk of errors when testing hypotheses.

The growth of Big Idea 5

Big Idea 5

Evaluating an estimator involves considering bias, precision, and the sampling method.

In grade 7, students informally assess sample-to-sample variation on the basis of random samples obtained through simulations. In high school, students work within a more explicit framework of estimation that clearly differentiates between sample and population and allows them to understand other measures of an estimator's performance, such as bias and standard error. Students' understanding of both bias and standard error are extensions of their earlier understanding of mean and standard deviation, applied to sampling distributions. (Note that the standard deviation is itself a more formal refinement of the mean absolute deviation measure of variability used in middle school.)

At the high school level, students expand their understanding and skill, acknowledging the risk of error when inferring values for population parameters and considering informal ways of measuring this error. Through interaction with an applet, students can see how often a particular confidence interval will miss the population parameter, and through this experience they can develop an appreciation for how often incorrect decisions might be made.

Students whose statistical reasoning continues to grow after grade 12 will discover that the assessment and quantification of this risk play an even more important role in the development and evaluation of statistical methods. For example, in regression, measuring the impact of omitting or adding variables to a model, and how well complex models fit the data, is a major focus of much of applied statistics.

Also embedded in Big Idea 5 is an understanding of the implications of sample size. Experiences in grades 6–8 mark the beginning of students' understanding of the role that sample size plays in inference, as students use data collected from random samples and simulations to make informal inferences about populations.

In high school, students' work in estimating population parameters makes the effect of sample size explicit. Students' informal ideas about the role of sample size often contain misconceptions, such as the mistaken idea (also common among adults) that small samples are more effective with small populations than with large populations (Garfield 2002). Students entering high school often have an intuitive notion that "more is better," but through their experiences in grades 9–12 they develop some sense of the costs of large sample sizes and the diminishing returns in increased precision.

Although high school students learn these ideas through metaphor and simulation, students who continue to develop their statistical understanding in later years will consider the mathematical foundations that allow for more formal proofs. The concepts in Big Idea 5 are especially important for statistical literacy since, for example, pundits regularly—and incorrectly—disparage public opinion polls for using sample sizes that are small relative to the population size.

Conclusion

In this chapter, we have seen how students' informal to semiformal to formal reasoning develops in light of each of the big ideas articulated in chapter 1. In chapter 3, we use an understanding of the big ideas and a developmental view of statistical reasoning to address learning, teaching, and assessment challenges.

Chapter 3

Challenges: Learning, Teaching, and Assessing

One of the biggest challenges for mathematics teachers in the coming years will be managing the changing role of statistics in the middle school and high school curriculum. As the Common Core State Standards for Mathematics (CCSSM) (Common Core State Standards Initiative [CCSSI] 2010) are implemented in most states, high school mathematics teachers will find themselves in the dual role of mathematics teacher and statistics teacher. Because the Common Core State Standards place heavy emphasis on conceptual understanding, embracing this new dual role will require a concerted effort as teachers strive to develop students' conceptual understanding of and ability to communicate statistical ideas.

The task of developing and assessing students' computational fluency in statistics—for example, their ability to compute the standard deviation, construct a graphical display, or fit a least squares line—is relatively easy compared with the challenge of developing and assessing their statistical reasoning. For this reason, we have chosen in this chapter to focus on developing students' statistical reasoning—something that teachers may be less familiar and less comfortable with than developing and assessing their procedural fluency.

Going beyond Computational Fluency

Becoming statistically literate in today's world requires that students understand key concepts of statistics, such as the importance of study design, the effect of the method of data collection on the type of conclusions that are reasonable, the fundamental ideas of sampling variability, and the logic involved in using data from a sample or from an experiment to draw conclusions about a population or about treatment effects in an experiment. You will recognize these as recurring themes in the big ideas and essential understandings of chapter 1. As with mathematical reasoning, statistical reasoning must be nurtured and developed over time. We offer four suggestions for enriching the teaching of statistics and the development of students' statistical reasoning:

- Look beyond procedural fluency.
- Ask good statistics questions.
- Give students the chance to practice "talking statistics."
- Provide authentic assessments and meaningful feedback.

We briefly consider each of these teaching "tips" in the four paragraphs that follow.

Teaching Tip 1
Look beyond procedural fluency.

These days, the ability to compute is of relatively little value by itself. Being able to compute a standard deviation or calculate a confidence interval estimate of a population mean is not of much use without knowing what these computations reveal about a population of interest. It is important to take students beyond the mechanics and be explicit about the logic of statistical procedures and the interpretation of results. Students need to demonstrate more than an ability to compute. When students ask why they have received only a small part of full credit for correct computations in response to a problem, just tell them that if the ability to compute is all they have to offer, they can be replaced by a $69 calculator!

Teaching Tip 2
Ask good statistics questions.

In a talk at the Joint Mathematics Meetings in 2010, Allan Rossman said that the key to being a great statistics teacher is to ask good questions. So what makes a good statistics question? Although that question itself probably has many answers, one that seems especially relevant in the context of developing students' statistical reasoning is that good questions should incorporate a conceptual component or an interpretation component—or both. This is an important idea that we develop more fully later in this chapter.

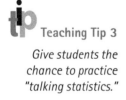

Teaching Tip 3
Give students the chance to practice "talking statistics."

Data analysis has a communication component as well as a mathematical component. The abilities to interpret and communicate results in context are not innate—they are skills that develop slowly and with practice. Although it might be a bit painful to listen to or read through students' initial attempts at explaining why they selected a particular method or how they have interpreted results from a data analysis, opportunities to practice, coupled with meaningful and constructive feedback, are essential. By providing these, teachers can successfully guide students' development in "talking statistics." Attention to developing this aspect of statistical reasoning does take patience, but it can be very rewarding.

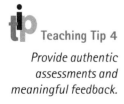

Teaching Tip 4
Provide authentic assessments and meaningful feedback.

If statistical reasoning is to be valued, it is important that we assess both conceptual understanding and the ability to interpret results in a meaningful way. Because assessment is such an important

Challenges: Learning, Teaching, and Assessing

topic, and because many mathematics teachers don't have much experience with assessment in the area of statistics, we also revisit this subject in much greater detail later in this chapter.

Producing Good Statistics Questions

The goal of looking beyond procedural fluency has implications for the kinds of questions that we want students to consider. To see how this idea plays out in relation to the questions that we ask, let's consider how we might ask a question about the standard deviation—a content topic encompassed in Big Idea 1 and central to all the big ideas in this volume. Reflect 3.1 directs your attention to question 1 below:

> **Question 1 (Not Very Good!)**
>
> What is the value of the standard deviation of the following 10 numbers?
>
> 2009 2015 2002 1979 2032 1991 2016 2030 2001 1990

> **Reflect 3.1**
>
> We have labeled question 1 "not very good." Why do you think we chose to describe it in this way?

Question 1 assesses only computational skills and is not a very good statistics question. In fact, some would argue that it is not a statistics question at all because it gives the numbers without any context. One way to improve this question would be to add a context that gives meaning to the numbers and to the standard deviation, as does question 2, which is the focus of Reflect 3.2:

> **Question 2 (Better, But Not Much!)**
>
> The data below are the lifetimes (in hours) for 10 light bulbs from a new brand that your school is considering for use in the football stadium light fixtures:
>
> 2009 2015 2002 1979 2032 1991 2016 2030 2001 1990
>
> What is the value of the standard deviation of the 10 lifetimes?

> **Reflect 3.2**
>
> We have labeled question 2 as "better, but not much" in comparison with question 1. In what way is it better? Why did we choose to describe it as "not much better"?

Question 2 is better than question 1 because it includes a context. We now know that the first value, 2009, represents a light bulb with a lifetime of 2009 hours. But the question is not much better, because it still focuses only on computation of the standard deviation. The information provided by the context is irrelevant to what students are asked to do with the data. Reflect 3.3 seeks a comparison of question 2 with question 3 below:

Question 3 (Better...)

The data below are the lifetimes (in hours) for 10 light bulbs from a new brand that your school is considering for use in the football stadium light fixtures:

2009 2015 2002 1979 2032 1991 2016 2030 2001 1990

a. What is the value of the standard deviation of the lifetimes for the 10 light bulbs from the new brand?

b. The standard deviation for the lifetimes of bulbs from the brand currently in use is 40 hours. What does the standard deviation that you computed for the sample of light bulbs from the new brand tell you about how this brand might compare with the old brand?

> **Reflect 3.3**
>
> What does part *b* of question 3 add that was missing from question 2?

Although part *a* of question 3 is still computational, part *b* asks students to go beyond computation to provide an interpretation of the standard deviation that makes sense in the context of the data. Question 4, which appears at the top of the next page, takes question 3 a step further. The new question includes a computational component and also has a part that asks students to interpret in context, as they do in responding to question 3. But the addition of part *c* enriches the question by asking students to answer a question of

Question 4 (Even Better)

The data below are the lifetimes (in hours) for 10 light bulbs from a new brand that your school is considering for use in the football stadium light fixtures:

2009 2015 2002 1979 2032 1991 2016 2030 2001 1990

a. What is the value of the standard deviation of the lifetimes for the 10 light bulbs from the new brand?

b. The standard deviation for the lifetimes of bulbs from the brand currently in use is 40 hours. What does the standard deviation that you computed for the sample of light bulbs from the new brand tell you about how this brand might compare with the old brand?

c. Replacing stadium light bulbs is difficult and requires special equipment. Because of this, rather than replace individual bulbs as they burn out, the school plans to replace *all* the stadium light bulbs as soon as one burns out. The mean lifetime is 2000 hours for both the current brand and the new brand under consideration, and the cost of the two brands is the same. Would you recommend that the school stay with the current brand or change to the new brand? Explain your reasoning.

interest that requires them to demonstrate conceptual understanding of the standard deviation and what it measures. Question 4 has a relevant context, requires interpretation, and also requires that students demonstrate a level of conceptual understanding. These are the characteristics of a good statistics question. Reflect 3.4 provides an opportunity to evaluate an assessment question.

Reflect 3.4

Consider the following assessment item. What is good about this item? How might it be improved?

> A random sample of size 40 was used to compute a 90% confidence interval for the population mean, resulting in (74.6, 82.9). Would a 95% confidence interval computed from the same sample data be wider than, narrower than, or the same width as the given 90% confidence interval?

The assessment item in Reflect 3.4 does explore students' understanding of an important statistical concept—the relationship between the width of a confidence interval and the confidence level. This item could be made better in a variety of ways, including the following:

- An interesting context could be added, and students could also be asked to provide an interpretation of the interval in context.

- The question could be modified to ask students for an explanation of *why* they responded that the new interval would be wider, narrower, or the same width. Adding the simple question, "Why do you think this?" would make this a better item for formative assessment. Then, if students' responses were incorrect, it would be easier to identify and address the errors in their thinking.

When you are developing questions that you plan to use as a basis for class discussion or for assessment, consider evaluating your questions by asking yourself the following questions:

- Does the question provide a meaningful context? (If not, it may not be a "statistics" question at all!)

- Is the context relevant to the problem—that is, do students need to make use of the context?

- Does the question have an interpretation component?

- Does the question have an aspect that requires students to demonstrate an understanding of the relevant concepts?

- If the context, interpretation, or conceptual component is missing, is there a way to improve the question by incorporating one or all of these components?

Of course, not every question that you use to guide classroom discussion or assign as a homework problem will do all these things. But finding as many ways as you can to give students opportunities to interpret results and communicate their understanding of fundamental concepts can be very beneficial in developing students' statistical reasoning abilities.

Assessing Interpretation and Conceptual Understanding

Assessing computational fluency is relatively easy because students' responses to computational questions can generally be determined to be correct, mostly correct (with some minor arithmetic errors),

Challenges: Learning, Teaching, and Assessing

or incorrect, and student errors are usually easy to identify and address. By contrast, assessing students' abilities to interpret the results of a data analysis or their understanding of important statistical concepts is more difficult. One possibility is to use a holistic rubric to classify students' responses. You might consider adapting one or both of the four-point rubrics in figure 3.1 to meet your own needs.

4-Point Rubrics	
For questions that address communication and interpretation	**For questions that address conceptual understanding**
3 The interpretation is appropriate, complete, and well communicated.	3 Demonstrates an understanding of the relevant concept that is clearly communicated.
2 The interpretation is appropriate and complete, but it is not communicated well.	2 Demonstrates some understanding of the relevant concept, but the understanding is not communicated clearly.
1 The interpretation includes appropriate statements that demonstrate some understanding, but it is incomplete or not quite correct.	1 Demonstrates only limited understanding of the relevant concept.
0 No interpretation provided, or it is incorrect and/or inappropriate.	0 Demonstrates little or no understanding of the relevant concept.

Fig. 3.1. Two 4-point rubrics using a scale of 0–3 for assessing students' understanding of statistical concepts

To see how student responses might be classified according to the rubric for questions that address conceptual understanding, let's revisit part c of question 4 discussed above:

c. Replacing stadium light bulbs is difficult and requires special equipment. Because of this, rather than replace individual bulbs as they burn out, the school plans to replace *all* the stadium light bulbs as soon as one burns out. The mean lifetime is 2000 hours for both the current brand and the new brand under consideration, and the cost of the two brands is the same. Would you recommend that the school stay with the current brand or change to the new brand? Explain your reasoning.

A conceptual understanding of variability is required to answer this question. Reflect 3.5 explores an assessment of a possible student response.

Reflect 3.5

Using the rubric for questions that address conceptual understanding (see fig. 3.1), a teacher gave a score of 0 to the response below:

> I don't think it makes any difference which brand the school uses. The mean lifetime is the same for both brands, so the school may as well stay with the old brand. There is no reason to change to the new brand.

Why do you think the teacher scored the response in this way?

The teacher gave the student response in Reflect 3.5 a score of 0 because it fails to recognize the role that variability plays in a decision about which brand to recommend. The response does not demonstrate any understanding of variability. Reflect 3.6 presents another possible student response.

Reflect 3.6

The teacher gave a score of 1 to the student response below:

> The means for the two brands are the same, but the standard deviation of the brand currently in use is larger, so there will be some longer lifetimes with the current brand than there would be with the new brand. We should stay with the current brand.

In what way is this response better than the one in Reflect 3.5?

The response in Reflect 3.6 acknowledges that the different standard deviations should be considered, but the recommendation not to switch to the new brand shows that the student does not understand how to interpret the standard deviation *in the context of the question posed*. Consequently, the response demonstrates only limited understanding of variability and the use of the standard deviation to describe variability. Reflect 3.7 presents yet another possible student response for consideration.

Reflect 3.7

How would you score the following student response?

> *The standard deviation of the new brand is less than the standard deviation of the current brand. There appears to be less variability in the lifetimes of the new brand, so the lifetimes of the new brand will tend to be more similar. We should switch to the new brand.*

A teacher might give a score of 2 to the response in Reflect 3.7. The response recognizes that less variability is desirable, although it fails to acknowledge that this may be an advantage only if the mean lifetime of the new brand is equal to or greater than the mean for the current brand. The response also does not clearly explain why a small degree of variability is desirable in this context. Compare this response with the more complete one in figure 3.2. This response would merit a score of 3. Of course, students will not write responses like this one without a great deal of practice!

A response meriting a score of 3:

There appears to be less variability in the lifetimes of the new brand, so the lifetimes of the new brand will tend to be more similar. Both the current brand and the new brand have a mean lifetime of 2000 hours. Because the current brand is more variable, the first bulb to burn out is likely to be quite a bit before 2000 hours. For the new brand, lifetimes cluster more tightly around 2000 hours, and so the first bulb to burn out would probably be closer to 2000 hours. So, all the bulbs will get replaced sooner for the current brand compared to the new brand. The cost will be lower if we switch to the new brand.

Fig. 3.2. A student response meriting 3 points on a 4-point scoring rubric (from 0 to 3 points)

In providing students with meaningful feedback, it is useful to think about how you might help them develop deeper understanding and express their thinking more clearly—that is, how you might help them "move up the rubric." Reflect 3.8 offers an opportunity to revisit the responses in Reflect 3.5, 3.6, and 3.7 with this goal in mind.

> **Reflect 3.8**
>
> Consider the student responses in Reflect 3.5, 3.6, and 3.7. In each case, what feedback would you provide that might help the student produce a higher-scoring response?

The students who wrote the responses that received scores of 0 and 1 are not adequately recognizing the role of variability. You might provide the following feedback:

> Take a look at the following two sets of lifetimes (in hours), both of which have the same mean lifetime:
>
> New brand: 1998   1999   2000   2001   2002
>
> Current brand: 1980   1990   2000   2010   2020
>
> How do these two data sets differ? When would all five bulbs be replaced in the case of the new brand? When would all five bulbs be replaced in the case of the current brand? How would you use your answers to these questions to reconsider your answer to the light bulb question?

This feedback might help students develop more insight into the role of variability in the context of the recommendation of a brand, thereby enabling them to provide a better response to this question—and perhaps to similar questions in the future.

The student who wrote the response that received a score of 2 probably understands the concept of variability and how it relates to the brand choice in this context. But this student needs to be more explicit in explaining the reasoning behind the response. To help this student, you might take the same feedback approach of providing two hypothetical data sets that would challenge the explanation. For example, you might present the following sets of bulb lifetimes (in hours):

New brand:	1900	1901	1902	1903	1904
Current brand:	1980	1990	2000	2010	2020

Considering these two data sets would show that less variability, by itself, is not enough of a basis for the response, which also needs to appeal to the fact that the mean lifetime and cost are the same for the two brands.

The Data Analysis Process—Culminating Assessments

One of the goals of the middle school and high school statistics curriculum is for students to develop an understanding of data analysis as a process with four essential steps:

1. Formulating a question that can be answered by collecting, analyzing, and interpreting data
2. Developing a thoughtful plan for data collection
3. Using graphical and numerical statistical methods to analyze the resulting data
4. Interpreting and communicating results

This process is first introduced in the middle grades, and then the various steps in the process are developed more deeply in grades 9–11. By the time students reach grade 11, they should be able to put all the steps together to demonstrate an understanding of the process as a whole. One way to develop and evaluate students' understanding of this process is through the use of individual and group projects. Resources on incorporating and assessing statistics projects are available on the Web; see, for example, "Learn More about Projects" at https://apps3.cehd.umn.edu/artist/projects.html. In addition, each year the American Statistical Association sponsors a poster and project competition for K–12 students, and sample projects as well as tips for creating a good statistical poster or project can be found at the competition website: http://www.amstat.org/education/posterprojects/index.cfm. This site also contains sample scoring rubrics that could be useful in assessing students' work on both projects and posters.

Conclusion

In this chapter, we have tried to provide some ideas about what makes a good statistics question and some suggestions about how teachers might assess responses to good statistics questions. Teachers who "own" the big ideas and essential understandings presented in chapter 1 and the insights into the development of statistical reasoning outlined in chapter 2 are well positioned to act on these recommendations. For teachers who are new to teaching statistics, the task of absorbing and acting on all these ideas may seem a bit overwhelming at first, but just as students get better at answering good statistical questions, this task too becomes easier with experience!

Appendix 1
Glossary of Statistical Terms

more4U
This glossary is also available in printable electronic form at www.nctm.org/more4u.

alternative hypothesis: the competing hypothesis in hypothesis testing, denoted by H_a and sometimes called the *research hypothesis*. The alternative hypothesis is determined by the statistical question of interest. (See also null hypothesis.)

approximate sampling distribution: the distribution of the observed values of a sample statistic (such as the sample mean) for many random samples of a given size, selected from the population. Used to describe how the value of the statistic varies from sample to sample, the approximate sampling distribution differs from the sampling distribution in that it is constructed by simulating the sampling process.

bias: a systematic difference between an estimate of a parameter and the true value of that parameter. Bias can arise from a poor sampling design, for example, if some segment of the population is underrepresented in the sample, and can also be a result of the estimator if the mean of the estimator's sampling distribution differs from the true value of the parameter.

bivariate data: observations in which two responses are measured for each individual, such as diameter and height for each tree, or height and arm span for each person, in a sample.

categorical variable: a variable whose possible values are not numerical, such as gender or political affiliation.

conditional probability: the probability of an event, *given that* some other event has occurred.

confidence interval: a range of values considered to be plausible for the population parameter, computed by taking a measure of the variability of an estimator (standard error) into account.

confidence level: a percentage that quantifies the chance that a future confidence interval will include the population parameter. Thus, the confidence level is associated with the method used to construct an interval estimate.

confounding factor (also called *confounding variable*): a variable that is associated with both the response variable and the primary independent variable of interest in an experiment or observational study. Confounding factors provide possible alternate explanations for a difference between treatment groups and limit ability to draw cause-and-effect conclusions.

controlled experiment: an investigation in which researchers impose a treatment or experience on the subjects.

deviation from the mean: the difference between a particular data value and the mean of the data values, computed as $x_i - \bar{x}$.

estimator: a statistic used to estimate a population parameter.

exponential distribution: a characteristic asymmetric distribution for a continuous numerical variable, illustrated below:

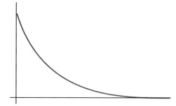

interquartile range: the distance between the 25th and 75th percentiles of the values in a data set, used as a measure of the variability of the data.

margin of error: the maximum likely mistake in prediction—that is, the most that the value of a sample statistic is likely to differ from the actual value of the population parameter for a predetermined confidence level.

mean: for sample data, the arithmetic average—that is, the sum of the data values divided by the number of data values—used as a measure of the center of a distribution. Interpreted as the "balancing point" of a distribution, the mean can be used to describe a "typical" value in a sample or population.

median: for sample data, the numeric value that separates the larger 50% of the data values from the smaller 50%, used as a measure of the center of a distribution.

normal distribution: A characteristic bell-shaped and symmetric distribution for a continuous numerical variable, illustrated below:

A normal distribution can be completely specified by its mean value, which locates its center, and its standard deviation, which measures its spread about the mean.

Appendix 1

null hypothesis: the hypothesis that is initially assumed to be true in a hypothesis test, denoted by H_0. (See also **alternative hypothesis**.)

numerical variable: a variable whose observed values are numeric.

observational study: an investigation in which the researcher observes characteristics of one or more than one sample to draw inferences about one or more than one population.

parameter: a numeric measure that describes a characteristic of the population—for example, the mean or the standard deviation.

population: the entire set of individuals or objects or values under investigation.

population distribution: the distribution of the values of some variable for the entire population and describing its individual-to-individual variability. In most real-life contexts, the population distribution is unknown.

precision: a measure of sample-to-sample variability of a statistic, typically described by the standard error.

p-value: the probability of observing data as extreme as, or more extreme than, those that were observed in the sample, given that the null hypothesis is true.

random assignment: the process of assigning subjects to experimental groups by using a chance mechanism.

random number table: a chart in which the digits 0 through 9 appear equally often in a random order, sometimes used to facilitate random selection or random assignment.

random sample: a sample composed of individuals selected from the population by using a chance mechanism. This term is often used to designate a **simple random sample**.

random selection: the process of selecting individuals for a sample by using a chance mechanism that ensures that every individual in the population has the same chance of being selected.

sample: a subset of a population, chosen for further study.

sample distribution: the distribution of the values of some variable for a sample selected from the population and describing individual-to-individual variability within the sample.

sampling distribution: The distribution of the values of a sample statistic, such as the sample mean, for all possible samples of a given size that can be selected from the population. The sampling distribution provides a description of how the value of the statistic varies from sample to sample.

sampling variability: the sample-to-sample variation resulting from the random selection process.

significance level: the conditional probability that the null hypothesis will be rejected, given that it is true. Usually, the value of the significance level is determined prior to performing a hypothesis test, and common choices for significance levels are .05, .10, and .01.

simple random sample: a subset of the population chosen in such a way that every sample of the same size is equally likely to be chosen, achieved by selecting individuals at random.

simulation: an imitation of random behavior created by using a mechanism that accurately reflects the situation.

standard deviation: a measure of variability, based on deviations from the mean.

standard deviation (population): a measure of a typical deviation from the mean when all values in a population are known, often denoted by σ and computed by using the formula

$$\sigma = \sqrt{\frac{\sum (x - \mu)^2}{n}}.$$

standard deviation (sample): a measure of a typical deviation from the mean for all data values in a sample, often denoted by s and computed by using the formula

$$s = \sqrt{\frac{\sum (x - \bar{x})^2}{n-1}}.$$

standard error: a measure of sample-to-sample variability of a statistic; the standard deviation of the sampling distribution of the statistic.

statistic: a numeric measure calculated from sample data, such as the mean or the standard deviation.

statistical inference: a branch of statistics that involves drawing conclusions about a population based on sample data.

stratified random sample: a sample drawn by using a sampling scheme that incorporates the use of supplemental information to divide the population into mutually exclusive subgroups, called *strata*.

treatment: a condition imposed on a subject to study its effect in an experiment. The experimenter assigns treatments at random to evaluate whether they differ in their effect on a response variable.

Appendix 1

trend: a systematic pattern, often of increasing or decreasing behavior, observed between two numerical variables.

unbiased estimator: a statistic that, on average, equals the value of the parameter.

univariate data: observations of a single characteristic, such as height.

This resource list is also available in printable electronic form at www.nctm.org/more4u.

Appendix 2
Resources for Teachers

The resources below have been selected to provide teachers with supplemental material to enhance their understanding of both the mechanics and the concepts of statistics, as well as to provide them with activities that they can use directly with their students in high school mathematics and statistics classes. General resources appear below, followed by specific resources targeted to requirements of the Common Core State Standards for Mathematics related to courses in algebra 1, geometry, and algebra 2. Some of the specific course resources identified below come from the sources that are listed as "general resources," whereas others are from specific books or journal articles.

General Resources

Textbooks, Journal Articles, and Activity Books

Agresti, Alan, and Christine Franklin. *Statistics: The Art and Science of Learning from Data*. 2nd ed. Upper Saddle River, N.J.: Pearson, 2009.

Beseler, Susan. "The Three-Point Shoot-Out: The Logic of Hypothesis Testing." *Mathematics Teacher* 99 (April 2006): 582–87.

Bock, David E., Paul F. Velleman, and Richard D. De Veaux. *Stats: Modeling the World*. 3rd ed. Boston: Pearson, 2010.

Bright, George W., Wallece Brewer, Kay McClain, and Edward S. Mooney. *Navigating through Data Analysis in Grades 6–8*. Principles and Standards for School Mathematics Navigations Series. Reston, Va.: National Council of Teachers of Mathematics, 2003.

Bright, George W., Dargan Frierson Jr., James E. Tarr, and Cynthia Thomas. *Navigating through Probability in Grades 6–8*. Principles and Standards for School Mathematics Navigations Series. Reston, Va.: National Council of Teachers of Mathematics, 2003.

Burrill, Gail F., and Portia C. Elliott, eds. *Thinking and Reasoning with Data and Chance*. Sixty-eighth Yearbook of the National Council of Teachers of Mathematics (NCTM). Reston, Va.: NCTM, 2006.

Appendix 2

Burrill, Gail, Christine A. Franklin, Landy Godbold, and Linda J. Young. *Navigating through Data Analysis in Grades 9–12. Principles and Standards for School Mathematics* Navigations Series. Reston, Va.: National Council of Teachers of Mathematics, 2003.

Cobb, George W., and David S. Moore. "Mathematics, Statistics, and Teaching." *American Mathematical Monthly* 104 (November 1997): 801–23.

Devlin, Thomas F. "Why Aren't They Called Probability Intervals?" *Mathematics Teacher* 101 (May 2008): 647–51.

Franklin, Christine, Gary Kader, Denise Mewborn, Jerry Moreno, Roxy Peck, Mike Perry, and Richard Scheaffer. *Guidelines for Assessment and Instruction in Statistics Education (GAISE) Report: A Pre-K–12 Curriculum Framework.* Alexandria, Va.: American Statistical Association, 2007.

Gelman, Andrew, and Deborah Nolan. *Teaching Statistics: A Bag of Tricks*. New York: Oxford University Press, 2002.

Gould, Robert, and Colleen Ryan. *Introductory Statistics: Exploring the World through Data*. Boston: Pearson, 2012.

Kader, Gary, and Tim Jacobbe. *Developing Essential Understanding of Statistics for Teaching Mathematics in Grades 6–8*. Essential Understanding Series. Reston, Va.: NCTM, 2013.

Peck, Roxy, George Casella, George Cobb, Roger Hoerl, Deborah Nolan, Robert Starbuck, and Hal Stern. *Statistics: A Guide to the Unknown*. 4th ed. Boston: Cengage Learning, in partnership with the American Statistical Association, 2006.

Peck, Roxy, Chris Olsen, and Jay L. Devore. *Introduction to Statistics and Data Analysis*. 4th ed. Boston: Cengage Learning, 2012.

Peck, Roxy, and Daren Starnes. *Making Sense of Statistical Studies*. Alexandria, Va.: American Statistical Association, 2009.

Peters, Susan A. "Engaging with the Art and Science of Statistics." *Mathematics Teacher* 103 (March 2010): 496–503.

Rossman, Allan J., Beth L. Chance, and J. Barr von Oehsen. *Workshop Statistics: Discovery with Data and the Graphing* Calculator. 3rd ed. Indianapolis: John Wiley & Sons, 2008.

Salsburg, David. *The Lady Tasting Tea: How Statistics Revolutionized Science in the Twentieth Century*. New York: W.H. Freeman, 2001.

Scheaffer, Richard L., Ann Watkins, Jeffrey Witmer, and Mrudulla Gnanadesikan. *Activity-Based Statistics*. 2nd ed., revised by Tim Erickson. Indianapolis: John Wiley & Sons, 2004.

Shaughnessy, J. Michael, Gloria Barrett, Rick Billstein, Henry A. Kranendonk, and Roxy Peck. *Navigating through Probability in Grades 9–12*. Principles and Standards for School Mathematics Navigations Series. Reston, Va.: National Council of Teachers of Mathematics, 2004.

Shaughnessy, J. Michael, and Beth Chance. *Statistical Questions from the Classroom.* Reston, Va.: National Council of Teachers of Mathematics, 2005.

Shaughnessy, J. Michael, Beth Chance, and Henry Kranendonk. *Focus in High School Mathematics: Reasoning and Sense Making in Statistics and Probability.* Reston, Va: National Council of Teachers of Mathematics, 2009.

Starnes, Daren S., Daniel S. Yates, and David S. Moore. *The Practice of Statistics*. 4th ed. New York: W.H. Freeman, 2012.

Vickers, Andrew. *What Is a p-Value Anyway? 34 Stories to Help You Actually Understand Statistics.* Boston: Pearson, 2010.

Watkins, Ann E., Richard L. Scheaffer, and George W. Cobb. *Statistics in Action: Understanding a World of Data.* 2nd ed. Emeryville, Calif.: Key Curriculum Press, 2008.

Online Resources

American Statistical Association. *Journal of Statistics Education.* http://amstat.org/publications/jse

American Statistical Association. Webinars. http://amstat.org/education/webinars/index.cfm

Consortium for the Advancement of Undergraduate Statistics Education. Webinars. http://www.causeweb.org/webinar

Consortium for the Advancement of Undergraduate Statistics Education. Chance News. http://www.causeweb.org/wiki/chance/index.php/Main_Page

International Statistics Literacy Project. http://www.stat.auckland.ac.nz/~iase/islp/priteach

National Council of Teachers of Mathematics. Illuminations. http://illuminations.nctm.org/WebResourceList.aspx?Ref=2&Std=4&Grd=0

North Carolina School of Science and Mathematics. Statistics Institutes. http://www.ncssm.edu/courses/math/Stat_Inst/links_to_all_stats_institutes.htm

Statistics Education Web. http://amstat.org/education/stew/index.cfm

Statistics Teacher Network. http://amstat.org/education/stn

Appendix 2

Course-Specific Resources

Algebra 1 Resources

Bohan, Jim. "Using Regression to Connect Algebra to the Real World." In *Thinking and Reasoning with Data and Chance*, Sixty-eighth Yearbook of the National Council of Teachers of Mathematics (NCTM), edited by Gail F. Burrill, pp. 195–208. Reston, Va.: NCTM, 2006.

Bright, George W., Wallece Brewer, Kay McClain, and Edward S. Mooney. *Navigating through Data Analysis in Grades 6–8*. Principles and Standards for School Mathematics Navigations Series. Reston, Va.: National Council of Teachers of Mathematics, 2003.

Burke, Maurice J., and Ted R. Hodgson. "Using Technology to Optimize and Generalize: The Least-Squares Line." *Mathematics Teacher* 101 (September 2007): 102–7.

Canada, Daniel L. "The Known Mix: A Taste of Variation." *Mathematics Teacher* 102 (November 2008): 286–91.

Erickson, Tim. "A Pretty Good Fit." *Mathematics Teacher* 102 (November 2008): 256–62.

Kader, Gary D., and Christine A. Franklin. "The Evolution of Pearson's Correlation Coefficient." *Mathematics Teacher* 102 (November 2008): 292–99.

Shaughnessy, J. Michael, and Maxine Pfannkuch. "How Faithful Is Old Faithful? Statistical Thinking: A Story of Variation and Prediction." *Mathematics Teacher* 95 (April 2002): 252–59.

Wilson, David C. "The Median-Median Line." *Mathematics Teacher* 104 (November 2010): 262–67.

Geometry Resources

Albert, Jim. "Interpreting Probabilities and Teaching the Subjective Viewpoint." In *Thinking and Reasoning with Data and Chance*, Sixty-eighth Yearbook of the National Council of Teachers of Mathematics (NCTM), edited by Gail F. Burrill, pp. 417–33. Reston, Va.: NCTM, 2006.

Bright, George W., Wallece Brewer, Kay McClain, and Edward S. Mooney. *Navigating through Data Analysis in Grades 6–8*. Principles and Standards for School Mathematics Navigations Series. Reston, Va.: National Council of Teachers of Mathematics, 2003.

Danielson, Christopher, and Eric Jenson. "Probability in Practice: The Case of Turkey Bingo." *Mathematics Teacher* 102 (November 2008): 248–54.

Rubel, Laurie H. "Students' Probabilistic Thinking Revealed: The Case of Coin Tosses." In *Thinking and Reasoning with Data and Chance*, Sixty-eighth Yearbook of the National Council of Teachers of Mathematics (NCTM), edited by Gail F. Burrill, pp. 49–59. Reston, Va.: NCTM, 2006.

Scheaffer, Richard L., Ann Watkins, Jeffrey Witmer, and Mrudulla Gnanadesikan. *Activity-Based Statistics*. 2nd ed., revised by Tim Erickson. Indianapolis: John Wiley & Sons, 2004. (Specifically, see the activities "What Is Random Behavior," "What's the Chance?" and "Dueling Dice.")

Shaughnessy, J. Michael, Gloria Barrett, Rick Billstein, Henry A. Kranendonk, and Roxy Peck. *Navigating through Probability in Grades 9–12. Principles and Standards for School Mathematics* Navigations Series. Reston, Va.: National Council of Teachers of Mathematics, 2004.

Algebra 2 Resources

Burrill, Gail, Christine A. Franklin, Landy Godbold, and Linda J. Young. *Navigating through Data Analysis in Grades 9–12. Principles and Standards for School Mathematics* Navigations Series. Reston, Va.: National Council of Teachers of Mathematics, 2003.

Koban, Lori, and Erin McNeils. "Fantasy Baseball with a Statistical Twist." *Mathematics Teacher* 102 (November 2008): 264–71.

Peck, Roxy, and Daren Starnes. *Making Sense of Statistical Studies*. Alexandria, Va.: American Statistical Association, 2009.

Scheaffer, Richard L., Ann Watkins, Jeffrey Witmer, and Mrudulla Gnanadesikan. *Activity-Based Statistics*. 2nd ed., revised by Tim Erickson. Indianapolis: John Wiley & Sons, 2004. (Specifically, see the activities "Random Rectangles," "Streaky Behavior," "Spinning Pennies," and "How Accurate Are the Polls?")

Tarr, James E., Hollylynne Stohl Lee, and Robin L. Rider. "When Data and Chance Collide: Drawing Inferences from Empirical Data." In *Thinking and Reasoning with Data and Chance*, Sixty-eighth Yearbook of the National Council of Teachers of Mathematics (NCTM), edited by Gail F. Burrill, pp. 139–49. Reston, Va.: NCTM, 2006.

Teague, Daniel J. "Experimental Design: Learning to Manage Variability." In *Thinking and Reasoning with Data and Chance*, Sixty-eighth Yearbook of the National Council of Teachers of Mathematics (NCTM), edited by Gail F. Burrill, pp. 151–69. Reston, Va.: NCTM, 2006.

References

Ben-Zvi, Dani, and Abraham Arcavi. "Junior High School Students' Construction of Global Views of Data and Data Representations." *Educational Studies in Mathematics* 45 (January 2001): 35–65.

Box, George, William Hunter, and J. Stuart Hunter. *Statistics for Experimenters: Design, Innovation, and Discovery.* 2nd ed. Hoboken, N.J.: Wiley-Interscience, 2005.

CBS News. "Duct Tape Therapy." *Healthwatch.* February 11, 2009. http://www.cbsnews.com/stories/2002/10/14/health/main525523.shtml.

Chance, Beth, Robert delMas, and Joan Garfield. "Reasoning about Sampling Distributions." In *The Challenge of Developing Statistical Literacy*, edited by Dani Ben-Zvi and Joan Garfield, pp. 295–323. New York: Springer, 2004.

Common Core State Standards Initiative (CCSSI). *Common Core State Standards for Mathematics. Common Core State Standards* (College- and Career-Readiness Standards and K–12 Standards in English Language Arts and Math). Washington, D.C.: National Governors Association Center for Best Practices and the Council of Chief State School Officers, 2010. http://www.corestandards.org.

delMas, Robert, and Yan Liu. "Exploring Students' Conceptions of the Standard Deviation." *Statistics Education Research Journal* 4 (May 2005): 55–82.

Derry, Sharon, Joel Levin, Helen Osana, Melanie Jones, and Michael Peterson. "Fostering Students' Statistical and Scientific Thinking: Lessons Learned from an Innovative College Course." *American Educational Research Journal* 37 (Fall 2000): 747–73.

Devore, Jay. *Probability and Statistics for Engineering and the Sciences.* 8th ed. Boston: Cengage Learning, 2012.

Diaconis, Persi. "Magical Thinking in the Analysis of Scientific Data." *Annals of the New York Academy of Sciences* 364 (June 1981): 236–44.

Dumanovsky, Tamara, Cathy Nonas, Christina Huang, Lynn Silver, and Mary Bassett. "What People Buy from Fast-Food Restaurants: Caloric Content and Menu Item Selection, New York City 2007." *Obesity* 17 (July 2009): 1369–74.

El-Hassan, H., K. McKeown, and F. A. Muller. "Clinical Trial: Music Reduces Anxiety Levels in Patients Attending for Endoscopy." *Alimentary Pharmacology & Therapy* 30 (October 2009): 718–24.

Franklin, Christine, Gary Kader, Denise Mewborn, Jerry Moreno, Roxy Peck, Mike Perry, and Richard Scheaffer. *Guidelines for Assessment and Instruction in Statistics Education (GAISE*

Report): A Pre-K–12 Curriculum Framework. Alexandria, Va.: American Statistical Association, 2007. http://www.amstat.org /education/gaise.

Freedman, David, Robert Pisani, and Roger Purves. *Statistics*. 4th ed. New York: W.W. Norton & Company, 2007.

Garfield, Joan. "The Challenge of Developing Statistical Reasoning." *Journal of Statistics Education* 10 (November 2002). http:// www.amstat.org/publications/jse/v10n3/garfield.html.

Hancock, Chris, James Kaput, and Lynn Goldsmith. "Authentic Inquiry with Data: Critical Barriers to Classroom Implementation." *Educational Psychologist* 27, no. 3 (1992): 337–64.

Kader, Gary, and Tim Jacobbe. *Developing Essential Understanding of Statistics for Teaching Mathematics in Grades 6–8*. Essential Understanding Series. Reston, Va.: National Council of Teachers of Mathematics, 2013.

Lee, Wesley, Mamtha Balasubramaniam, Russell L. Deter, Sonia S. Hassan, Francesca Gotsch, Juan Pedro Kusanovic, Luis Flavio Goncalves, and Roberto Romero. "Fetal Growth Parameters and Birth Weight: Their Relationship to Newborn Infant Body Composition." *Ultrasound in Obstetrics and Gynecology* 33 (April 2009): 441–46.

Lipson, Kay. "The Role of Computer Based Technology in Developing Understanding of the Concept of Sampling Distribution." In *Developing a Statistically Literate Society: Proceedings of the Sixth International Conference on Teaching Statistics*. 2002. http://www.stat.auckland.ac.nz/~iase /publications/1/6c1_lips.pdf.

Lohr, Sharon. *Sampling: Design and Analysis*. 2nd ed. Belmont, Calif.: Duxbury Press, 2009.

Makar, Katie, and Andee Rubin. "A Framework for Thinking about Informal Statistical Inference." *Statistics Education Research Journal* 8 (May 2009): 82–105.

National Council of Teachers of Mathematics (NCTM). *Curriculum and Evaluation Standards for School Mathematics*. Reston, Va.: NCTM, 1989.

———. *Principles and Standards for School Mathematics*. Reston, Va.: NCTM, 2000.

———. *Curriculum Focal Points for Prekindergarten through Grade 8 Mathematics: A Quest for Coherence*. Reston, Va.: NCTM, 2006.

———. *Focus in High School Mathematics: Reasoning and Sense Making*. Reston, Va.: NCTM, 2009.

Zieffler, Andrew, Joan Garfield, Robert DelMas, and Chris Reading. "A Framework to Support Research on Informal Inferential Reasoning." *Statistics Education Research Journal* 7 (November 2008): 40–58.

Titles in the Essential Understanding Series

The Essential Understanding Series gives teachers the deep understanding that they need to teach challenging topics in mathematics. Students encounter such topics across the pre-K–grade 12 curriculum, and teachers who understand the related big ideas can give maximum support as students develop their own understanding and make vital connections.

Developing Essential Understanding of—

Number and Numeration for Teaching Mathematics in Prekindergarten–Grade 2
    ISBN 978-0-87353-629-5   Stock No. 13492

Addition and Subtraction for Teaching Mathematics in Prekindergarten–Grade 2
    ISBN 978-0-87353-664-6   Stock No. 13792

Rational Numbers for Teaching Mathematics in Grades 3–5
    ISBN 978-0-87353-630-1   Stock No. 13493

Algebraic Thinking for Teaching Mathematics in Grades 3–5
    ISBN 978-0-87353-668-4   Stock No. 13796

Multiplication and Division for Teaching Mathematics in Grades 3–5
    ISBN 978-0-87353-667-7   Stock No. 13795

Ratios, Proportions, and Proportional Reasoning for Teaching Mathematics in Grades 6–8
    ISBN 978-0-87353-622-6   Stock No. 13482

Expressions, Equations, and Functions for Teaching Mathematics in Grades 6–8
    ISBN 978-0-87353-670-7   Stock No. 13798

Geometry for Teaching Mathematics in Grades 6–8
    ISBN 978-0-87353-691-2   Stock No. 14122

Statistics for Teaching Mathematics in Grades 6–8
    ISBN 978-0-87353-672-1   Stock No. 13800

Functions for Teaching Mathematics in Grades 9–12
    ISBN 978-0-87353-623-3   Stock No. 13483

Geometry for Teaching Mathematics in Grades 9–12
    ISBN 978-0-87353-692-9   Stock No. 14123

Proof and Proving for Teaching Mathematics in Grades 9–12
    ISBN: 978-0-87353-676-9   Stock No. 13803

Statistics for Teaching Mathematics in Grades 9–12
    ISBN 978-0-87353-676-9   Stock No. 13804

Mathematical Reasoning for Teaching Mathematics in Prekindergarten–Grade 8
    ISBN 978-0-87353-666-0   Stock No. 13794

Forthcoming:

Geometry for Teaching Mathematics in Prekindergarten–Grade 2

Geometric Shapes and Solids for Teaching Mathematics in Grades 3–5

Visit www.nctm.org/catalog for details and ordering information.